Success in Science

A Manual for Excellence in Science Education

Bradley & Paige Hudson

Success in Science

A Manual for Excellence in Science Education

First Edition 2012

Email: info@elementalscience.com

ISBN: 978-1-935614-21-0

Cover Design by Bradley & Paige Hudson

Printed In USA For World Wide Distribution

For more copies write to :
Elemental Science
175 W Monroe St #131
Wytheville VA 24382
info@elementalscience.com

Success in Science: Table of Contents

Introduction

The Purpose of this Book

Science is a subject near and dear to our hearts. Brad loved the hands-on research opportunities that he had in high school so much that he went on to pursue a double major in the sciences as well as a masters in biology, while Paige was so impacted by her chemistry courses in high school that she decided to go on to pursue a bachelor's degree in biochemistry. Our experiences with science have been overwhelmingly positive, but the sad reality is that, for many people this is not the case. Most of the population cringes when you mention words like chemistry or physics. For years, science has been looked at as a lofty subject that few really understand.

The truth is that science is the explanation for what we see going on around us every day. Science explains why the ice in our glass melts quicker on a hot day than a cold one. Science gives us the reasons for how our body works and why the weather is what it is. Science surrounds us daily and it helps us to really understand what is going on in our environment.

The typical reason for studying science in high school is because every major university requires that a student has taken several of these courses before he can apply for admissions. But beyond the academic reasons, science education can be beneficial for every student. We believe that there are two main purposes for teaching science.

First, science will give the student a better awareness of the world around him. Science is all around us; it's in the metamorphosis of the butterfly, in the changing colors of the leaves in autumn and in the apple falling off the tree. The fact that you can walk upright and digest your food can all be explained through science. When you teach science properly, the student learns the why's behind the things that take place every day around him, resulting in a deeper appreciation of life.

The second purpose for teaching science is because it helps to train the brain to think logically. The scientific method is the formula that all scientists learn to use when approaching a problem. The scientist has a question, researches about that question's topic, predicts the answer to his question, does a test and then analyzes what he has found to determine what the answer to his question could be. This method is a logical and thorough process that the scientist will use over and over again. Being familiar with the scientific method will help to train the brain of the student to approach any question or problem in a logical manner.

In this book, we will provide you with a framework for teaching science to the various grade levels as well as lay a foundation for why we feel you need to place an importance on science education. Our desire is to see all students excel in their study of the fields of science. Our hope is that this book will give you the tools you need to see your student shine as he studies those fields.

Part 1: The Foundation of Science Education

1

The Most Important Concept

The study of science helps to develop the student's ability to think logically and critically. By analyzing problems and determining how to follow the scientific method, the student gains confidence along with a foundation for logical thinking. According to Carl A Rotter...

> *"While teachers can do nothing to increase a student's mental capacity they can modify their instructional strategies to make concepts easier to comprehend. This may be accomplished through the use of concrete models, illustrations and diagrams and hands on experiences. A student's cognitive developmental growth will increase through exposure to activities requiring them to reason formally. Students, who use integrated science process skills during science activities, increase the level of their cognitive development."*

Another study by Süleyman Yaman found that science education trains people to "discover, explore, make right decisions, solve problems and continuously learn".

We have seen that teaching science well promotes excellence in our students, but to truly understand science, one must be familiar with a single key concept: the scientific method. No high school student should graduate without a firm grasp on this concept; simply memorizing the method is not enough. The student needs to have used the scientific method over and over in experiments until it has become a natural habit. It will take years for a student to fully etch the scientific method into his mind. So, the scientific method is something that you need to begin teaching from the very start.

What is the Scientific Method?

In a nutshell, the scientific method teaches the brain to logically examine and process all the information it receives. It requires that one observes and tests before making a statement of fact. It is the main method scientists use when asking and answering questions. The main steps of the scientific method are...

1. Ask a Question
2. Research the Topic
3. Formulate a Hypothesis
4. Test with Experimentation
5. Record and Analyze Observations and Results
6. Draw a Conclusion

Using the scientific method will teach the student to look at all the evidence before making a statement of fact. It sounds like a lofty idea, but in reality it is an integral part of science education. If we want our students to be prepared for higher education science, they must be comfortable with this most fundamental process.

Though this all sounds intimidating, it's really not. You are simply teaching the student to take the time to discover the answer to a given problem by using the knowledge he has as well as the things he observes and measures during an experiment. The scientific method is a simple, yet logical process that follows the same steps every time.

The Steps in Detail

In the remainder of this chapter we will look at each of the steps in the scientific method in more detail in hopes that you will become more comfortable with teaching the process to your students.

Step 1: Ask a Question

The scientific method begins when a scientist observes an occurrence that makes him wonder what is happening. He then creates a question relating to what he has perceived. When crafting the inquiry, the scientist makes sure that the question is worded in such a way that he will be able to measure whether or not he has obtained the answer. Good questions begin with how, what, when, who, which, why or where.

For example, let's say the student is fascinated by the growth of the plants that he observes, but he doesn't know why this is happening. So a question he could formulate would be, "Why do plants grow?", but this will be time-consuming to measure, so, you will want to help him narrow down the question. Some options are: "How does the lack of sunlight affect the growth of house plants?" or "Which soil is best for house plants to be grown in?" Each of these questions is more specific, making them far easier to measure.

Step 2: Research the Topic

Next, the scientist researches about the topic from the question so that he will have some background knowledge about the subject. This keeps him from repeating mistakes that have been made in the past, but also gives him a basis for formulating his hypothesis. It is very hard to predict what is going to happen in an experiment without knowing something about the principles at work.

The student from the example in step 1 would start by reading about plants and researching how they grow. He can begin with the encyclopedias or reference books that he already has on hand. Next, the student should look at the library for any books relating to plants. Finally, he can search the internet for scientific articles relating to his topic. The student will glean a lot of information from his research, so you will need to assist him in determining which information is useful for answering his question and which material can be tossed.

Step 3: Formulate a Hypothesis

In step 3, the scientist formulates his hypothesis, which is a fancy word for an educated guess about the answer to his question. The hypothesis must always be able to be measured as well as provide the answer to the original question that was asked. Hypotheses are normally simple "if-then" statements that are not more than one sentence long.

For example, the student has chosen the question, "How does sunlight affect the growth of house plants?" He has done his research and found out that sunlight is transformed into energy by the chlorophyll found in the plant cells. So, his hypothesis could be, "*If a plant receives less sunlight, then it will stop growing.*" or "*If a plant receives more sunlight, then the plant will grow more.*"

Step 4: Test with Experimentation

The next step in the scientific method is for the scientist to develop an experiment that will test whether his hypothesis is true or false. It is important for the test to be fair, so the scientist will only change one variable at a time and he always has a control group. He generally has more than one sample in each group so that the findings will be reliable. The scientist may also find that several experiments are necessary to thoroughly prove whether his hypothesis is correct or not.

In our example, the student will need to design an experiment that tests whether or not a plant will grow if the presence of sunlight is changed. This is a relatively easy hypothesis to test, since it only contains one independent variable, the amount of sunlight. His experiment could have nine house plants that are allowed to grow on a shelf in a relatively sunny room. After 5 days, 3 of the house plants can be moved into a completely dark room, 3 of the house plants can be moved into the full sun on a window sill and the other 3 house plants are left on the shelf for 5 more days. He will need to water each one as necessary, so that the amount of water does not become another variable in his experiment.

A Word About Variables

Each experiment seeks to test a variable, which is an event or factor that you are trying to measure. There are two main types of variables, independent variables and dependent variables, which can be found in every experiment. The independent variable is the factor that is controlled or changed by the scientist performing the experiment, such as the amount of sunlight the plant receives in the above experiment. The dependent variable is the factor being tested in the experiment, like the amount that the plant grows in the experiment that we have been discussing. The dependent variable is what the scientist measures to determine the effect of the changes to the independent variable. In other words, the dependent variable depends upon the independent variable.

It's also important to mention controlled variables. A controlled variable is a factor that is not being examined in the experiment. The scientist will keep the controlled variable constant so that its effect on the experiment will be minimized. The amount of water would be an example of a controlled variable in the experiment that we have been discussing.

Step 5: Record and Analyze Observations and Results

During the experiment, the scientist will record all his observations and measure his results. Observations are a record of the things he has seen happen in the experiment, while results are the specific and measurable data that he has collected during the testing. Once the experiment is complete, he will analyze the observations and results to see if his hypothesis was true or false.

So let's say that the student we have been discussing has kept a journal of his observations during the experiment. His journal contained entries like:

"Day 7: The plant on the window sill is green and tall. The plant on the shelf is green, but a little shorter than the plant on the window sill. The plant in the dark is turning yellow and has not grown in days."

He has also measured how much the plants grew each day and then plotted each of his measurements on a graph that looks like this:

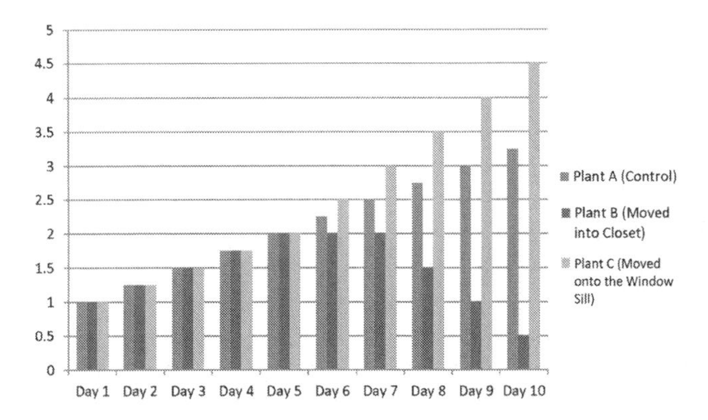

Now the student can easily see from his journal entries and from the graph he created that his hypothesis was true. He has found the answer to his question.

Step 6: Draw a Conclusion

Finally, the scientist can use what he has discovered to make a statement about whether or not his hypothesis was true. This statement will communicate his results to other scientists and hopefully answer his original question. The scientist may find that his hypothesis was false or that his experiment design did not really answer his question. If this is the case, he will formulate a new hypothesis and begin the process again until he are able to answer his question.

In our sample experiment, the student saw that the plant in the dark room stopped growing and began to shrink, while the plant on the window sill grew faster than the one on the shelf. He can easily theorize that, "*The more sunlight a plant has, the better it will grow.*" However, more testing is needed to see if other independent variables play a role in plant growth before he can make a statement of fact about the relationship of plant growth and the presence of sunlight.

A Quick Word about Theory vs. Fact

If you noticed in the last paragraph, we stated that the student could theorize, but he could not make a statement of fact. The student is able to make an educated prediction that an increase in exposure to sunlight will cause an increase in the plant's growth. However, the student cannot make a statement of absolute truth because he has not examined all the factors that affect the plant's growth and how those factors relate to an increased presence of sunlight.

So what is the difference between a theory and a fact? To answer this question, we must examine the origins of the words themselves. The word theory comes from the ancient Greek word *theoria*, which means "a looking at, viewing or beholding". In science, a theory is an analytical tool used for understanding, explaining or predicting cause for a certain subject matter. The word "fact" comes from the Latin word *factum*, which means "a thing done or performed". In science, a fact is an objective truth that can be seen in nature or confirmed through repeated experiments.

So, we can say that theories are meant to be tested by experimentation and observation to determine if they are fact, while facts are truths that can be verified through repeatable experimentation or by real-time observations.

Conclusion

Whether you are creating a theory or proving a fact, the scientific method is the tool that every scientist uses to determine the findings. The scientific method is integral to understanding science, but it has benefits that are useful beyond the study of how the world works. It is a technique that trains the student how to answer a question in a logical manner. It also teaches the student to analyze and process the information he is receiving. The scientific method teaches the brain to logically examine and process all the information it receives. This is why we believe this process is the most important concept in science education.

In the next section of this book, we will lay out our plan for teaching science throughout the preschool, elementary, middle school and high school years.

Part 2: What Science Education Should Look Like

2
The Preschool Years

The preschool student is learning daily about his environment. He is constantly absorbing information about the world around him through hands-on experiences. He enjoys seeing how things work and loves being introduced to new things. The preschooler is more than ready and willing to learn, but his motor muscles aren't quite ready for all the writing that formal education entails.

Why Teach Science to a Preschooler?

Typically, the preschooler is taught the basics, such as colors, the alphabet and the numbers 1-20 through simple worksheets. We also provide him with structured play, such as a kitchen set or a dress up station. We make sure that he has time to build his motor skills through creating art and exploring music. However, all too often, we neglect to introduce the youngest student to the wonder of science because we think it is too difficult of a subject for him to grasp. While some concepts in science will go way over a preschooler's head, we can introduce him to the subject as a whole by presenting him with the way that things work in his environment.

Chapter 2: The Preschool Years

The preschooler is naturally wired to be curious, and thus, he is fully prepared to learn about science. These early years are a good time to introduce him to the way things work in his environment, because an early introduction to the subject will create an interest that you can build upon once he reaches the elementary years. By showing him the miracle of the scientific processes going on around him, you are constructing a basis for future learning.

Your Goal

The goal for preschool science is simple:

1. To introduce the student to the world around him.

The preschool student is a completely blank slate, so during these years your goal will be to introduce him to various concepts and ideas found in science through a hands-on approach. This will help him to build a basic framework, or bucket, that he can fill during the elementary years.

The Components

There are four basic components to preschool science education that will help you to accomplish your goal. They are...

1. The Weekly Topic
2. Practical Projects
3. Read-Alouds
4. Coordinating Activities

What you accomplish each week will vary because the student's interest will vary. Some weeks a preschooler will want to spend every waking minute learning something; some weeks he will only want to spend 5 minutes on educational topics. So look at the following components as a buffet of ideas that you can use to introduce the student to the world of science rather than a list of things to do each week.

Remember that science during the preschool years needs to be very hands-on and teacher directed. During these years, science should also be strictly enjoyable for the student. If he does not enjoy reading books about science (or any of the other components), don't force him to do so, as this will be counter-productive to your goal. There will be plenty of years for him to learn what he needs to in the not so distant future.

A Closer Look At The Components

The four components for your preschool science buffet are the weekly topic, hands-on projects, read-alouds and coordinating activities. A good science curriculum for the preschool years will give you options for each of the following components.

The Weekly Topic

The main purpose of having a weekly topic is to create a focus for your studies for the week. Once you choose your weekly topic, create a main idea that relates to the topic. This main idea should put science into words that a preschooler understands, such as, "*Rain is water falling from clouds in the sky.*" Next, you will determine a way to introduce the topic to the student. These should be simple explanations, demonstrations and/or guided observations that the student will understand.

When you give the introduction, it should include the main idea that you will seek to emphasize throughout the week. You can have the student color a page related to the topic or have him copy the main idea and illustrate it on a separate sheet of paper. If he has questions during your introduction, answer them, but guide him to stay on the topic you are studying as the preschool mind tends to wonder. The introduction should only take 5 to 15 minutes because of the preschoolers short attention span.

Practical Projects

The practical projects component of your buffet should include simple scientific demonstrations or nature studies. The demonstrations should be designed to help the student see the science of his environment in action, while nature studies should be designed to aid the student in learning about the world around him through discovery and observation. You can choose to include one or both of these options in your week.

Scientific demonstrations can come from books like *Science Play*, *Mudpies & Magnets* or *More Mudpies & Magnets*. The goal of these demonstrations is to display scientific phenomenon for the student, which will allow him to discover more about the world around him. Don't expect preschoolers to be able to predict the outcome of or to draw abstract conclusions about these scientific demonstrations. Instead, allow him to observe what is happening and tell what he has learned, no matter how simple it may seem to you.

Nature studies can be as simple as taking a walk with purpose in your neighborhood. A walk with purpose means that as you walk you are looking for the topic being studied in the nature around you. So if you are studying flowers that week, on your walk you would point out the different flowers that you encounter as well as discuss the diverse shapes, sizes and colors that you see. A more thorough explanation on what nature study is can be found in the Appendix.

Read-Alouds

During the preschool years, the student usually loves to be read to and science is a good topic to explore through books at this age. The local or school library will typically have a non-fiction section that will be packed with science-related material. Simply look up the topic for the week, skim the available books for appropriateness and take home what the student will enjoy listening to. You can also choose fiction books that are related to the topic you are studying to add even more interest. Either way, make sure to choose books that you know will interest the student.

Coordinating Activities

Coordinating activities are meant to reinforce what the student is learning in science. You can use crafts or snacks that relate to the topic being studied for the week, but remember that the crafts should be relatively simple so that the student will not be exasperated by them. You can also incorporate simple finger plays or songs to enhance his learning. As you complete these projects, make sure to share the main idea for the week with the student, as it will help to reinforce what he has been studying.

What about a Science Notebook?

The science notebook can be a simple journal where the

student records what he has learned over the year. The bulk of the notebook should be pictures that the student has drawn, colored or pasted onto his pages. Most (if not all) of the writing in the notebook should be done by you, since his fine motor skills are not completely developed and writing can be a chore. The science notebook should be an enjoyable record of what he has learned instead of a mandatory task he has to complete. It can also be a place to store all the papers and projects that the student has completed in science for the year.

What Can It Look Like

The following is a sample week for preschool science on the topic of rain. We have included this illustration to give you an idea of what you can achieve in a week with a preschool-age student.

Weekly Topic: Rain

Main idea: Rain is water falling from clouds in the sky.

Introducing the Topic: Share the following 19th century nursery rhyme with the student:

Rain, Rain,
Go away;
Come again,
April day;
Little Johnny wants to play.

Say to the student: "Do you know what rain is? (*Answers will vary; praise him if he is correct, but guide him to the answer if he is not.*) Rain is water falling from clouds in the sky. Water evaporates from the lakes and oceans around us. It goes up into the air and forms clouds. When those clouds get really heavy, they let go of some of the water and it falls to the ground as rain. This week we are going to study rain."

When you are finished, have him color a coordinating page with the topic, like the one pictured. If he is comfortable with writing, you could have him copy the main idea into his science notebook instead.

Practical Projects

You can choose to do either a scientific demonstration or a nature study, or you can do both activities.

Scientific Demonstration: *Science Play* pg. 59, "Raindrop Landing Pad"

In this scientific demonstration the student will discover the shapes raindrops can make. This would be better to do on a rainy day, but if that is not possible you can simulate the rain with an eye dropper and some water.

Materials needed:
- ✓ Flour
- ✓ Salt
- ✓ Aluminum pan

After the student observes the shapes that the raindrops have made, have him trace those shapes into his science notebook.

Nature Study: Observing Rain

Go on a nature walk right after a rain shower. Have the student look for signs that it has rained, such as water on the grass and leaves, puddles on the sidewalk or steam rising from the road. Point out these signs to him if he doesn't see them on his own. Once you return home, allow the student to sketch what he wants in his science notebook if you are keeping one.

Read-Alouds

You could read any of the following books to the student for the week:

1. *Down Comes the Rain* (Let's-Read-And-Find... Science: Stage 2) by Franklyn Mansfield Branley and James Graham Hale
2. *The Rain Came Down* by David Shannon
3. *Rain* (Weather Series) by Marion Dane Bauer and John Wallace

Coordinating Activities

You can choose to do one or all of the following activities.

Activity: Measure the Rain

This activity is explained on pg. 59 of *Science Play*. You will need a clear jar, a ruler and a permanent pen.

Craft: Pitter Patter Paint

This activity is explained on pg. 59 of *Science Play*. You

will need tempera paint, paper and rain. If it is not raining on the day you do this craft, simply use a water sprayer to simulate the rain.

Snack: Orange Umbrellas

Cut an orange into thin round slices, and then slice each round in half for the top of your umbrella. Use pretzel sticks or thinly sliced apples for the handle of your umbrella. Eat and enjoy!

Conclusion

The preschool years are perfect for awakening the wonder of science in a student. The preceding goal and components we have laid out in this chapter will give you the tools you need as a teacher to do just that.

3
The Elementary Years

The elementary student is focused on learning the foundational basics, such as reading, spelling and simple mathematical operations. However, this does not mean that we should neglect the sciences during this time. These are the years when the student is still full of wonder and when he desires to know everything about everything. The elementary student peppers you with questions like, "*What's ____________?*" So through these years, you want to capitalize on his natural curiosity by showing him how much fun science can be. You also want him to become aware of the world around him through observing the things he sees every day.

The elementary years start when formal schooling begins, in first grade, and end when the student wants to know the why's and how's of what he observes. For some, this means that they end in the 4th grade, for others they end in the 5th and for some other students this phase ends earlier or later than that. Either way, it is crucial to begin teaching science during these years. The elementary student needs to learn that science can be enjoyable and to see it in action with his own eyes. He needs to engage with the subject as a whole, while learning some of the most basic facts of each discipline.

Your Goals

There are two goals for science education during the elementary years:

1. To create interest in the student for the learning of science.
2. To fill the student's mind with interesting (but basic) scientific information.

The elementary student is an empty bucket that is begging to be filled. He has a natural curiosity that is combined with a high capacity for retaining information. So, your goals for teaching science to him will play to these strengths while filling his bucket with scientific information. You can also use science to work on his basic skills of writing and reading at this stage.

The Components

Your approach to science during the elementary years will have three basic components and two optional ones. The basic components of elementary science are:

1. Scientific Demonstrations or Observations
2. Science-Oriented Books
3. Notebooking

The optional components are:

1. Multi-week Projects
2. Memorization

'The way you put these components together varies on the method that you choose, which we will share more about later in the chapter. For now, let's take a closer look at each of the components.

The Basic Components

The three basic components are scientific demonstrations, science-oriented books and notebooking. A good science curriculum for the elementary years must have each of these components.

Scientific Demonstrations or Observations

The purpose of doing scientific demonstrations or observations with the student at this level is threefold:

1. To work on his observation skills.
2. To introduce him to the scientific method.
3. To demonstrate science so that the student can discover the principles at work.

At this level you will be demonstrating scientific concepts for the student to observe. His job is to watch, absorb what is happening and file the information away for later use. However, if the student wants to participate and it is safe for him to do so, by all means allow him to, as long as you remain in control. As he progresses through the elementary years, the student will be able to participate more and more with these demonstrations. Just remember that you will always need to be there for him as a guide, controlling the flow of the scientific demonstration and being willing to answer any questions he may have along the way.

In each scientific demonstration, you are introducing them student to the scientific method by modeling the process for him. At this level, you will be skipping steps 2 and 3 of the scientific method because the student is not fully prepared for them. Dr. Susan Wise Bauer says in her lecture entitled *Science in the Classical Curriculum*, "You are not expecting them [the elementary-age student] to predict the results because they do not have a working knowledge of the principles at work to draw from." We agree that the elementary student should not be required to predict the results, i.e. formulate a hypothesis, for any scientific demonstration, because his "hypothesis" is not an educated guess, it's simply a guess. However, if the student is able to predict the outcome or enjoys guessing at the results, don't prevent him from doing so.

During the elementary years, we also recommend that you have the student fill out a basic lab report which will introduce him to the scientific method and prepare him for future years. His lab reports should include four sections:

- **Our Tools:** This section will list the materials that were used during the demonstration.
- **Our Method:** This section will contain the procedure for the demonstration in the student's words.
- **Our Outcome:** This section will contain what the student saw and record any data he has collected.
- **Our Insight:** The final section of his lab report will contain a sentence or two about what the student has learned from the demonstration. Ideally this will relate to the science being studied, but it's ok at this level for his sentences to be more superficial.

You can also use hands-on projects or nature studies for scientific observations at this level, as long as they relate to the topics you are studying. We do recommend that you do complete at least 4 scientific demonstrations along with lab reports for each year so that the student will be prepared for perfomring experiments during the middle school years.

Science-Oriented Books

At this stage the student is an empty bucket waiting to be filled with information and books are a wonderful way to do that. There are many children's encyclopedias, such the ones published by Usborne, Kingfisher and DK. Each of these publishers presents scientific information in an interesting way on the level of an elementary student. You can also choose to read living books that deal with science (*see the Appendix for a brief explanation of living books*). Either way, the books you choose to read need to be visually appealing, interesting and scientifically sound.

In the beginning of the elementary years, you will be reading these books to the student, but as his reading abilities increase, you can assign him books to read on his own. As he proceeds onto reading more complicated books, you can add in one or two books about scientists per year. These will help the student to engage with the face of science which will create an interest to learn more.

Notebooking

The purpose of the notebooking component is to verify that the student has placed at least one piece of information into his knowledge bucket. If you are not familiar with notebooking see the Appendix for a more thorough explanation of what it entails. Basically, notebooking is an extremely effective tool that will teach the student how to assimilate and release information. This is done by asking him more open-ended questions like "Was there something interesting that you found about what we just read?" We prefer to use notebooking at this stage because the student is more apt to engage with the material and therefore will be more likely to retain what he have been learning.

In the beginning you should not expect a lot of actual writing from the student. You can discuss his answers to your questions and then write them down for the student in his workbook. As he progresses through the elementary years, you can expect him to do more of his own writing for science, but don't force him to write beyond his ability, as pushing the elementary student in writing will lead to frustration with science and be contrary to your goals.

The Optional Components

If you find that the student really enjoys studying science, we suggest that you add one or both of the optional components to your science routine.

Multi-week Projects

Multi-week projects are a great way for the student to engage with the material he is learning over a longer period of time. These projects need to relate to what you are studying and reinforce what the student needs to know. For example, if you are studying a group of animals over several weeks, consider creating a chart in which the diets (carnivore, herbivore or omnivore) of each of the animals is displayed. This type of project will be a visual reminder of what the student has studied and will reinforce the concept of animal diet.

Memorization

Remember that the elementary student is an empty bucket that is begging to be filled and memory work is another tool that you can use to fill his knowledge banks with information that he can draw upon later. We recommend that you have the student memorize scientific poems, lists of facts or vocabulary that relate to what you are teaching him.

What the Student Needs to Learn During the Elementary Years

The elementary years are meant for sparking a student's interest in learning science, but there are some basic facts that he should be familiar with by the end of the period. Here is a list of them by discipline:

Biology

- know the difference between warm-blooded and cold-blooded animals
- know the types of animal diets (carnivore, herbivore, omnivore)
- be familiar with the types of animals (mammals, birds, reptiles, amphibians, fish and invertebrates) and a few basic facts about each (i.e., mammals-have lots of hair, feed their young with milk and are warm-blooded)

- know that humans are classified as mammals
- know that the human body contains different systems, each with a specific function
- be familiar with the basic parts of a plant (roots, stems, leaves, flower and seed) and what they do (i.e. roots-take up nutrients)

Earth Science

- be familiar with the natural cycles (i.e. be aware of what a natural cycle is and be familiar with the water cycle)
- know the 4 basic biomes (grassland, forest, desert, arctic) found around the world and a fact about each (i.e. The grasslands are characterized by an abundance of grasses.)
- know several of the more common weather formations (i.e. tornadoes, hurricanes and thunderstorms) as well as what volcanoes and earthquakes are
- be familiar with the types of rock (metamorphic, sedimentary, igneous)
- be familiar with what a fossil is

Astronomy

- know what is in our solar system (i.e. stars, asteroids, planets and so on)
- know the names of the planets

- be familiar with what stars and constellations are, plus a few of the major constellations (i.e. the Big Dipper)

Chemistry

- know what an atom, element and compound are
- be familiar with the periodic table (i.e. what it is and its purpose)
- know the 3 states of matter (i.e. gas, liquid and solid)

Physics

- have a very basic understanding of motion, gravity, friction, spin, balance (i.e. Gravity pulls things down towards the Earth.)
- know the colors of the rainbow
- be familiar with what magnets do (i.e. Magnets attract metal objects.)
- know several types of simple machines (i.e. lever, pulley and so on)
- know the major types of energy (light, sound, heat, potential, kinetic, chemical)

What Can It Look Like

During the elementary years there are many ways you can tackle teaching the necessary facts with the basic components. The method you choose will depend upon the interests of the student and your strengths as a teacher. No way is better than the other, as it really depends on the method that fits your style the most. The four elementary teaching methods that we are suggesting are Classic, Living Books, Nature Study and Unit Study. Each of these methods is explained with greater detail below. We have also included a sample week broken down into the components we explained above for each method so that you can see what the routine could look like.

The Classic Method*

The Classic approach to elementary science education is centered on scientific demonstrations that are enhanced with reference-book learning. The student will watch and participate in a demonstration each week about the topic. Then, he will read about the topic in an encyclopedia or textbook and write about what he has learned in his student notebook. If the time and interest allow, you can add additional projects or vocabulary to memorize. What follows is a sample week using the Classic method of elementary science instruction for the planet Venus.

Scientific Demonstration or Observation

Choose one of the scientific demonstrations from Janice Van Cleave's *Astronomy for Every Kid* such as "Hot Box" on pages 22-23. This demonstration will help the student see why Venus is so hot. Once he has completed the demonstration, have him fill out a simple lab report. A second grader's report might look like this:

- **Our Tools:** *2 thermometers, 1 jar with a lid*
- **Our Method:** *We put one thermometer in a jar and closed the lid. We recorded the temperatures. We set both*

* **Note:** The Classic method is loosely based on the principles of classical education. To learn more about this educational philosophy, we recommend reading *The Well-trained Mind: A Guide to Classical Education at Home* by Jessie Wise and Susan Wise Bauer.

thermometers outside and waited for 20 minutes.

- **Our Outcome:** *The temperature in the jar rose higher.*

	Plain Thermometer	Thermometer in the Jar
Initial Temperature	80°F	80°F
Final Temperature	93°F	104°F

- **Our Insight:** *The project was fun. Venus is the hottest planet because some of the heat gets trapped.*

Science-Oriented Books

Now read to the student (or if he is able have him read) about Venus out of an encyclopedia, such as Usborne's *First Encyclopedia of Space* or Kingfisher's *Discover Science Solar System.* Then, spend a few moments discussing what you just read. You can do this by asking leading questions that will pull out the most important information or by asking more broad questions that will help you to see what data the student has absorbed.

Notebooking

After you have completed the reading and discussion time, have the student tell you several things that he has learned about the planet Venus. Then, write (or have him write) those things down on a sheet of paper. A second grader might write something like this:

"Venus is covered by a gas called carbon dioxide. Venus has a lot of volcanoes. Venus is second from

the Sun. It is very hot."

Next, have him draw the planet or glue a picture of Venus above what was written. You could also have an older elementary student define atmosphere and add it to a glossary he has created in his student notebook.

Multi-week Project

Have the student add Venus to his own model of the solar system. He can create his model as a 3-D mobile, on a large sheet of paper or on the wall.

Memorization

Have the student work on memorizing the planets in our solar system or the definition of atmosphere.

The Living Books Method

The Living Books approach to elementary science education is centered on living books that are augmented with scientific demonstrations or observations. The student will read or be read to from a living book like a classic, such as *The Burgess Bird Book for Children* or a more modern option like *The Sassafras Science Adventures*. Then, he will write about what he has learned and do a related scientific demonstration or hands-on project with you. If the time and interest allow, you can add a non-fiction book related to the topic or do an additional activity. What follows is a sample week using the Living Books method of elementary science instruction for zoology (the study of animals).

Science-Oriented Books

Read Chapter 2 of *The Sassafras Science Adventures Volume 1 Zoology* summarized below:

> *The chapter opens with Blaine and Tracey arriving at their first stop in the African grasslands. They join Nicolas Mzuri and 4 other guests on a photo safari through Kenya. The group spots a pride of lions and Nicholas shares more about them. They also race several cheetahs on the hunt and their guide tells more about the animals. All the while, the narcoleptic Hanks will fall asleep and the comical couple, Fred and Pam, will make their characteristic blunders and mistakes. The chapter ends with the mysterious Man with No Eyebrows stealing the jeep, leaving the group stranded in Elephant Valley.*

This chapter introduces the student to the grasslands as well as lions and cheetahs. You can choose to break the chapter up into two readings or read it all at once. After you finish reading, spend a few moments discussing what you just read with the student. You can do this by asking leading questions that will pull out the most important information or by asking more broad questions that will help you to see what data the student has absorbed. If the student wants to read more, you can have him read about lions or cheetahs from either the DK's *Animal Encyclopedia* or Kingfisher's *First Animal Encyclopedia*, or choose another non-fiction book from the library.

Notebooking

After he has completed the reading and discussion time

with you, have the student tell you several things that he has learned about the lions and cheetahs. Then, write (or have him write) those things down on a sheet of paper. A typical first grader might say:

> *"Lions roar. Male lions have a mane. Cheetahs run very fast. They have spots."*

You could also have an older elementary student define food chain, grassland or mammal and add those definitions to a glossary he has created in his student notebook.

Scientific Demonstration or Observation

For this week, you can choose a hands-on activity, such as a foot race. You can highlight the fact that cheetahs are fast runners, but only over short distances. Then, you have a short foot race and declare the winner to be the cheetah of the group. You could also have the student complete a craft project related to lions or cheetahs.

Multi-week Project

Have the student create a shoebox diorama of the grasslands, adding in the lion and cheetah. He can continue to add to this project as he reads about the grasslands in the following chapter. Alternatively, have the student create an animal diet chart, where he places the lion and cheetah under the carnivore side. As he continues to read the book, he can place the various animals he studies under carnivore, herbivore or omnivore.

Memorization

Have the student begin to memorize the definitions of

food chain, grassland and mammal or the characteristics of a mammal, i.e. "Mammals are warm-blooded, feed their babies milk and have fur or hair covering most of their body".

The Nature Study Method

The Nature Study approach to elementary science education is led by finding science in nature and is enriched with reading from non-fiction books. For this method, the student will spend time outdoors studying the principles of science found in nature. Then, he will write about what he has learned and read a related non-fiction book on the subject. If the time and interest allow, you can do an additional project or craft. What follows is a sample week using the Nature Study method of elementary science instruction for rocks.

Scientific Demonstration or Observation

You will begin by taking a nature walk outside with the student, looking for various rocks to collect. Have him pick up as many different rocks as he can find. Once at home, use a rock identification manual or the Internet to classify the rocks he has collected.

Science-Oriented Books

Have the student read a book from the library about rocks. The following books are all good options:

- *If You Find a Rock* by Peggy Christian and Barbara Hirsch Lember
- *Rocks: Hard, Soft, Smooth, and Rough* by Rosinsky, Natalie M, John and Matthew

- *Let's Go Rock Collecting (Let's-Read-And-Find... Science: Stage 2)* by Roma Gans and Holly Keller
- *The Rock Factory: The Story About the Rock Cycle* by Bailey, Jacqui, Lilly and Matthew

After you finish reading, spend a few moments discussing what you just read. You can do this by asking leading questions that will pull out the most important information or by asking more broad questions that will help you to see what data the student has absorbed.

Notebooking

When you use the Nature Study method, it is best to have the student use a blank journal for his notebooking component. This way he will feel free to paste samples and pictures in there, to include drawings and to write about what he has learned. The nature journal should allow for the freedom to contain what the student wants to include. For this sample week, after he has completed the reading and discussion time with you, have the student tell you several things that he has learned about rocks. Then, write (or have him write) those things down in his nature journal A typical fourth grader might say:

> *"There are three different types of rock. Igneous rocks come from the heat in a volcano. Sedimentary rocks come from layers being squashed. Metamorphic rocks are rocks that change."*

You could also have an older elementary student define igneous rock, sedimentary rock, metamorphic rock or fossil and then add those definitions to a glossary he has created in the back of his nature journal.

Multi-week Project

Have the student create a "Rocks in My Area" display. You can have him take pictures of each of his samples and glue them into his nature journal or you can have him mount each of his samples on sturdy poster-board. Either way, the student needs to label each sample with the rock's name and where it was found.

Memorization

Have the student work on memorizing the following poem about rocks:

Types of Rocks

Igneous rocks are made from fire
They come from volcanoes that are sure to inspire
Sedimentary rock forms layer by layer
They are made of sand, mud and pebbles, not John Mayer
Metamorphic rock is formed by pressure and heat
They turn the rock into something really neat

The Unit Study Method

The Unit Study approach to elementary science education is based on a topic which the student explores by using the basic components. The student will spend time reading about the subject, narrating about what he has learned and completing a hands-on project or watching a scientific demonstration related to the topic. If the time and interest allow, you can read additional books on the subject or do an additional craft related to the subject with the student. What follows is a sample week using the Unit Study method of elementary science instruction for the alkaline earth metals of the periodic table.

Science-Oriented Books

Begin by reading to the student or having him read the section on alkaline earth metals on pages 22-30 in *Fizz, Bubble Flash*. When you are finished reading, discuss with the student what he has learned about the group as well as what he has learned about magnesium and calcium. You can do this by asking leading questions that will pull out the most important information or by asking more broad questions that will help you to see what data the student has absorbed.

Notebooking

After he has completed the reading and discussion time with you, have the student add a booklet on the alkaline earth metals to his lapbook on the periodic table (*for more information on what a lapbook is, see the Appendix*). The booklet should include the characteristics of the group, as well as information on magnesium and calcium. A typical third grade student should write:

- **Characteristics:** *Alkaline earth metals can be rocks. They react with air and water. They are very hard.*
- **Magnesium:** *Magnesium is in the food we eat. It is also found in chlorophyll.*
- **Calcium:** *Calcium was used by the Romans for teeth, chalk and a special type of cement.*

Scientific Demonstration or Observation

You can choose to do one of the activities suggested in *Fizz, Bubble, Flash*, such as "Get your Fill of Chlorophyll" on page 24 or "These Suds are Duds" on page 27. Discuss the lab report categories with the student and have him write his answers down on a simple lab report form. A third grader might write:

- **Our Tools:** *spinach, small cup alcohol, spoon, foil, scissors, paper towel, tape*
- **Our Method:** *We tore up the leaves, put them in a cup and covered them with alcohol. Then we used the spoon to mash them and let them sit for an hour. Then we dipped the coffee filter in the cup and taped it to the edge and watched for 60 minutes.*
- **Our Outcome:** *I saw orange and yellow.*
- **Our Insight:** *There are several colors in chlorophyll.*

Multi-week Project

Have the student add the alkaline earth metals to the periodic table on the wall that he is creating. This project can be done during the entire time you study the periodic table. As you place each group, talk about the groups' characteristics and point out the elements that you studied.

Memorization

Have the student work on memorizing the ten groups of the periodic table and where they are located.

Conclusion

Although there are many different methods for teaching elementary science, your goals remain the same. You want to spark the student's interest in science as well as fill his knowledge bucket with scientific information. The components and methods we have shared above will give you the tools to meet and exceed your goals for science during the elementary years.

4

The Middle School Years

During the middle school years, the student is beginning to apply the basic facts that he has mastered to the new concepts that he is learning. He has a strong desire to know why and is likely to question everything you are trying to teach him. The middle school student is assailing you with questions that begin with, "But why?" You want to capitalize on his need to know the reason as well as continue to feed him with information. During the middle school years you also want to emphasize inquiry-based methods as you teach the student how to organize and store information.

The middle school years begin when the student moves out of the elementary years and end when the student begins to take what he knows, analyze the whys behind it and then apply his conclusions to an unknown situation. For some students, this means they leave this stage by the end of the 8th grade, for others they leave by the end of 9th grade and for some other students this phase ends earlier or later than that. Either way, the middle school years are pivotal for preparing the student to receive the information he will learn in high school and beyond. The middle school student needs to explore the principles of science as well as learn the tools that he will utilize to organize the information he is receiving.

Your Goals

Your goals for science during the middle school years are three-fold:

1. To begin to train the student's brain to think analytically about the facts of science.
2. To familiarize the student with the basics of the scientific method through inquiry-based methods.
3. To continue to feed the student with information about the world around him.

We compare the middle school student to a bucket full of unorganized information that needs to be filed away and stored in a cabinet. This student has a strong desire to know why things work the way they do. So, when teaching science, you will be playing to his strengths while instructing him on how to make connections between the pieces of knowledge he deposited during the elementary years along with the information that he is currently learning. You can also use science to work on the basic skills of critical thinking and reading comprehension at this stage.

The Components

Your approach to science during the middle school years will have four basic components and two optional ones. The basic components are:

1. Hands-on Inquiry
2. Information

3. Writing
4. The Science Project

The optional components are:

1. Around the Web
2. Quizzes or Tests

How you put those components together will vary on the method you choose, which we will share more about later in the chapter. For now, let's look closer at each of the components.

The Basic Components

The four basic components are hands-on inquiry, information, writing and the science project. A good science curriculum for the middle school years must have each of the following components.

Hands-on Inquiry

The purpose of the hands-on inquiry component of the middle school years is three-fold:

1. To allow the student to experience real-life science.
2. To build the student's problem-solving skills.
3. To practice using the basics of the scientific method.

The middle school student can use nature study or laboratory experiments to fulfill this component.

Nature studies should focus on finding the answer to a scientific question in nature. Laboratory experiments should be more involved and more detailed at this stage than the scientific demonstrations done during the elementary years. Plus, your role will be shifting from demonstrator to supervisor. The student will move into performing the laboratory experiments on his own during these years and he will need to write an experiment report for each one. His experiment report should include the following sections:

1. **Title:** The title should be the question the student was attempting to answer or it should explain what he was observing.
2. **Hypothesis:** In this section, the student should share his prediction of the answer to the question posed in the experiment. If the experiment is an observation, he can share what he thinks he will see or skip this section of the experiment report.
3. **Materials:** The materials section should contain a list of what the student used to complete the experiment.
4. **Procedure:** In the procedure section, the student needs to write a step by step account of what he did during his experiment. This should be a summary of what was done in his own words so that someone reading it would understand what occurred.
5. **Observations and Results:** In this section the student will record what he observed during the experiment as well as any data that he has collected.
6. **Conclusion:** Finally, in the conclusion section the student will write whether or not his hypothesis was correct and any additional information that he has learned from the experiment. If his hypothesis was not correct, discuss why and have him include that on his experiment report.

For each experiment the student should run through the steps of the scientific method, with the exception of step 2.

The reason for skipping step 2 is because, at this point having to do research for an experiment every week would become cumbersome and could eventually lead to the student hating to do them. So, for now, you will still need to introduce each experiment by sharing a bit of background information on the topic. Then, take the time to go over the procedure with the student, making sure that he understands what he has to do. Once he has completed the experiment have him discuss his results with you.

Information

The information component is comprised of three interlocked pieces: books, vocabulary and diagrams. The middle school student still needs to be reading from an encyclopedia or textbook weekly as he needs to continue to build his knowledge base. During these years, he also needs to be discussing the material that he is reading with you on a more formal level, as this will help him to organize and store the information properly. Your discussion need to help him to pull out the key pieces of information, which means you will use comprehension-style questions, such as "How are plant and animal cells alike and how are they different?" By discussing what he has read, you are teaching the student to think critically about the passage as well as modeling how to pick out the important facts.

As you discuss the material, you will also want to highlight any unfamiliar vocabulary and have the student write those definitions into his student notebook. He can work on memorizing these terms as he progresses through his study of the topic. The student should also be drawing a diagram from what he has read, if possible. This drawing needs to be well labeled and represent the subject matter accurately so that it will give the student a more organized picture of the information that he has just learned. For example, it is important to know that a plant cell contains a cell wall, cell membrane, cytoplasm, vacuole, nucleus and chlorophyll, but it's far more effective for the student to have a mental picture of the parts of a plant cell and where he is located. Remember that your goal is not just to feed him with information, but also to help him to organize and think critically about what he is studying.

Writing

The purpose of the writing component is to teach the student how to process and organize information. You want him to be able to read a passage, pull out the main ideas and communicate them in his own words. Your discussion time with the student (part of the information component) has already prepared him for this task. Now, you are simply asking him to write a record of what he has learned and already discussed with you. You can assign short reports, note-taking, outlines or comprehension worksheets to fulfill the writing component. However, we much prefer to use outlining for the writing component as it will help the student to organize the material he is learning in a more logical manner.

Although writing is not always the favorite task of a middle school student, it is important that you do not skip this component. Having the student write out his thoughts in an organized manner will help him to shape the material in his informational filing cabinet. It will also give him yet another opportunity to interact with the material he is studying, which will serve to further cement the concepts into his mind.

The Science Project

Once a year, every middle school student should complete a science project. His project should work through the scientific method from start to finish on a basic level, meaning that his question should be relatively easy to answer (*see chapter 1 for a thorough explanation of the scientific method*). At this point, you need to be certain to thoroughly explain each step and coach the student through the entire process. He needs you to work alongside him as an advisor from the time he formulates his question until he polishes up his conclusion, so that he learns the process correctly from the beginning.

It is also important to have the student present his project to a group and answer related questions from them. This will reinforce what he has learned as well as help him to discern how to communicate what he knows. The best way to achieve this is to have the student participate in a Science Fair where his project will be judged, but if that's not possible, don't skip this component. The student can still present his project to his family or a group of his peers.

The Optional Components

If you find that the student really enjoys studying science, we suggest that you add one or both of the optional components to your science routine.

Around the Web

It's a good idea to have the middle school student gain some experience researching on the Internet. So for this optional component, you can have the student, under your supervision, search the Internet for websites, YouTube videos, virtual tours and activities that relate to what he is studying. You can also use this component as basis for a brief research report if you feel the student needs to work on that area.

Quizzes or Tests

During the middle school years it is not absolutely necessary that you give quizzes or tests to the student. However, if you want to familiarize him with test-taking skills, we suggest that you give quizzes or tests that will set the student up for success. You can do this just by testing material that has been thoroughly covered during class time. The questions can include matching vocabulary, true/false questions about the important concepts you have discussed as well as brief, short answer questions.

What the Student Needs to Learn During the Middle School Years

The middle school student is building upon his foundation of knowledge from the elementary years, but he is still not fully prepared for the math in science. Here's a list by discipline of key facts that he should know before leaving the middle school years.

Biology

- be familiar with the information listed in the elementary years chapter
- know the difference between animal and plant cells, along with several of the major parts (i.e. nucleus, cell

wall, cytoplasm and so on)

- know the units of classification (i.e. Kingdom, Phylum, Class, Order, Family, Genus, Species) along with the five major kingdoms (Moneran, Protist, Fungi, Plant and Animal)
- know the difference between a food chain and a food web as well as what the four major nutrient (natural) cycles are (i.e. nitrogen, phosphorus, carbon and water cycle)
- know the difference between fungi, simple plants, spore bearing plants and flowering plants (i.e. fungi reproduce using spores, while flowering plants have flowers which produce seeds) along with an example of each (i.e. Mushrooms are fungi.)
- be familiar with what deciduous and coniferous trees are along with an example of each (i.e. Oak trees are deciduous; pine trees are coniferous.)
- be familiar with the major groups of invertebrates (i.e. annelids, platyhelminthes, nematodes, cnidarians, mollusks, arthropods)
- be familiar with the major groups of vertebrates (i.e. fish, amphibians, reptiles, birds, mammals)
- be familiar with what migration is and how animals can defend themselves
- be familiar with each of the major systems of the human body along with what they do (i.e. The Integumentary System covers and protects the body.)

Earth Science

- be familiar with the information listed in the elementary years chapter
- be familiar with maps and how to read a map as well as know where the lines of longitude, lines of latitude, the equator and prime meridian are on the globe
- know the components of the Earth and where they are located (inner core, outer core, mantle, crust)
- know the seven continents (North America, South America, Europe, Asia, Africa, Antarctica and Australia or Oceania)
- know the layers of soil (topsoil, subsoil and bedrock)
- be familiar with the various types of rock weathering (i.e. freeze-thaw, chemical weathering and so on)
- be familiar with what causes volcanoes and earthquakes
- know the 8 major habitats and a characteristic of each (i.e. The Coniferous Forest is characterized by the predominance of coniferous trees, such as pines.)
- be familiar with what atmosphere and currents are as well as with the layers of the atmosphere (i.e. Sea Level, Thermosphere, Mesosphere, Stratosphere, Troposphere)
- be familiar with what causes weather and how it can be forecasted
- be familiar with the basic types of clouds (i.e. Cirrus,

Cirrostratus, Cirrocumulus, Altostratus, Altocumulus, Stratus, Stratocumulus, Nimbostratus, Cumulus, and Cumulonimbus)

- know the world's major oceans and seas along with where they are located (i.e. Arctic Ocean, Pacific Ocean, Atlantic Ocean, Indian Ocean, Southern Ocean, Mediterranean Sea, Red Sea, Black Sea, Caribbean Sea, Gulf of Mexico, Hudson Bay, Bering Sea, Tasman Sea, Coral Sea, Bay of Bengal, Arabian Sea, and North Sea)

Astronomy

- be familiar with the information listed in the elementary years chapter
- know what the universe is
- be familiar with the Big Bang Theory
- know the life cycle of a star
- know what a galaxy is, the name of our galaxy and the four types of galaxies (i.e. irregular, elliptical, spiral and barred spiral)
- be familiar with each of the planets in our solar system (i.e. Venus has a thick, poisonous atmosphere.)
- be familiar with what comets, asteroids and meteorites are
- know what a telescope is and what it is used for
- know what a satellite is and what it is used for

- be familiar with a few of the key astronomers (i.e. Copernicus, Galileo and so on)

Chemistry

- be familiar with the information listed in the elementary years chapter
- know what elements, atoms, ions are as well as how they relate to the periodic table
- know the groups of the periodic table
- know the difference between an elements and compounds, mixtures and solutions
- have a basic understanding of nomenclature for ions (i.e. SO_3 is sulfite and SO_4 is sulfate)
- be familiar with units of measurement and how to convert them (i.e. g to kg and so on)
- be familiar with what density is
- be familiar with chemical reactions, as well as what it means to be an exothermic (i.e. heat is released) and endothermic (i.e. heat is absorbed)
- have a basic understanding of how atoms bond together to form compounds and how that relates to the period table
- have a basic understanding of how gases, liquids and solids behave (i.e. how they behave relating to pressure, temperature, volume)

- know the difference between acids and bases as well as a basic understanding of pH and neutralization
- be familiar with what alcohols, hydrocarbons, and polymers are

Physics

- be familiar with the information listed in the elementary years chapter
- know the 3 Laws of Motion
- know what gravity, friction, and centripetal force are
- know what mass, speed, velocity and acceleration are and how they relate to motion
- know what a sound wave is and how amplitude affects sound
- know what a light wave is and how it behaves (i.e. reflection and refraction)
- have a basic understanding of thermodynamics (i.e. heat, work, power and so on)
- know the fact that energy is neither created nor destroyed, it is simply transferred from one type to another
- have a basic understanding of how electricity works as well as circuits and batteries
- be familiar with magnetism, how electricity can be used to create magnetism and what happens at the poles of a magnet

What Can It Look Like

The methods we are suggesting for teaching the basic components during the middle school years are similar in name to what we suggested during the elementary years. However, during this phase the student will be probing further into the basics as well as learning to organize the information he is studying. Even so, the method you choose will still depend upon the interests of the student and your strengths as a teacher.

The four middle school methods we are suggesting are Classic, Living Books, Nature Study and Unit Study, each of which are explained in greater detail below. As before, we have included a sample week broken down into the middle school components we explained above for each method so that you can see what each routine would look like. We have not included the science project in each of the sample weeks as it will remain the same for each method. Instead, we included a detailed explanation at the end of this chapter.

The Classic Method*

Similar to the elementary years, the Classic approach to middle school science education is centered on laboratory experiments that are enhanced with reference-book learning. The student will do an experiment each week, read about the topic in an encyclopedia or textbook and then write about what he has learned in his student notebook. During these years he will be doing his experiments with little help from you and his

* **Note:** The Classic method is loosely based on the principles of classical education. To learn more about this educational philosophy, we recommend reading *The Well-trained Mind: A Guide to Classical Education at Home* by Jessie Wise and Susan Wise Bauer.

reference books will be quite a bit more complex than it was during the elementary years. If time and interest allow during the week, you can have the student research the topic on the Internet. At some point during the year, you will take a break from his normal experiments to do a science project. What follows is a sample week using the Classic method of middle school science instruction about the sun.

Hands-On Inquiry

Laboratory Experiment: Do the sun's rays contain heat?

You will need a sunny day for this experiment along with 2 glass jars or 2 clear glasses, plastic wrap, 2 black tea bags, water and an instant read thermometer. Begin with introducing the topic to the student by saying:

> *"The sun is the closest star to the Earth. It is a giant ball of a constantly exploding gas called hydrogen. The surface of the sun is around 9,932°F, but the core is much hotter. It is the largest object in our Solar System. The sun's gravitational pull is so strong that all the planets in our Solar System revolve around it. The sun is important to the earth in many ways. This week we are going to look specifically at the sun's rays and how they affect the objects on our planet."*

Then go over the following procedure with the student and let him perform the experiment on his own.

1. *State your hypothesis (or answer to the question).*
2. *Fill each jar with the same amount of water. Take the temperature of the water in each jar. Then, place a tea bag in each jar and cover with the plastic wrap.*
3. *Set one of the jars outside in a sunny spot and leave one of*

the jars inside your house.

4. *After two hours, take the jar you left outside into the house. Then, measure the temperature of the water and observe any changes to the appearance of the water.*

The student should see that the jar that was outside in the sun has a greater temperature increase than the jar that was inside. He should also observe that the water in the outside jar is much darker than the water in the jar that was inside. In fact, he might see that the water in the inside jar did not noticeably change color. Discuss the reasons for this with the student by saying something like…

> *The jar that was left inside acts as a control because it did not receive any of the sun's rays. The jar that was outside in the sun received direct rays from the sun, which caused several changes to the water. The sun's rays carry energy in the form of heat which causes the temperature of the water to increase. This temperature increase also causes tea molecules to be released into the water, changing the color and taste of the water.*

Finally, have the student fill out an experiment report.

Information

Now, have the student read about the sun in an encyclopedia, such as the *Usborne Science Encyclopedia*, the *Kingfisher Science Encyclopedia* or *DK's Eyewitness Astronomy*. Discuss what he has read by asking him questions like:

1. *What is the sun?*
2. *What causes sunspots?*
3. *What is the chromosphere of the sun composed of?*

4. Briefly explain the expected life of the sun.

Then, have the student define photosphere, prominence, solar wind and sunspot and draw a diagram of the sun. His drawing should be fairly accurate, similar to the one pictured in his encyclopedia. He should label his diagram with the following parts of the sun: chromosphere, flare, facula, prominence, sunspot, hydrogen core and photosphere.

Writing

After the student has completed the reading and discussion time with you, have him outline the pages he read. A proficient sixth grader should write:

I. *The sun is a star that is made up of hot gas.*
II. *The sun contains sunspots caused by magnetic fields*
  A. *Sunspots can slow down the flow of heat within the sun.*
III. *The chromosphere of the sun is composed of hydrogen and helium.*
IV. *The sun will follow an expected path for its life.*
  A. *The sun was born from a nebula.*
  B. *Scientists believe that its life is halfway over.*
  C. *Once half of its hydrogen core has been converted into helium, the sun will become a red giant.*
  D. *The outer layer will shrink and the inner layers will collapse, turning the sun into a white dwarf.*
  E. *Then the sun will slowly fade out and die.*

Around the Web

Have the student research on the Internet for facts about the sun that he can use for a solar system fact book. He can continue this book as he studies the remaining planets in our solar system. Alternatively, have the student look at the following websites about the Sun (always preview the sites for appropriateness):

- **Virtual Tour of the Sun:** http://www.michielb.nl/sun/
- **Photos of the Sun:** http://science.nationalgeographic.com/science/photos/sun-gallery/

Quizzes or Tests

Once he has completed the unit, give the student a test on the solar system.

The Living Books Method

Similar to the elementary years, the Living Books approach to middle school science education is centered on living books that are augmented with laboratory experiments or nature studies related to the readings. The student will read from a living book, such as a classic like *The Wonder Book of Chemistry* or a more modern option like *The Sassafras Science Adventures.* Then, he will write about what he has learned and do a related experiment. If time and interest allow, you can have the student read about the topic in an encyclopedia or research the topic around the web. What follows is a sample week using the Living Books method of middle school science instruction for mixtures in chemistry.

Information

Begin by reading chapter 2 of *The Wonder Book of Chemistry* by Jeani Fabre. In this chapter, Uncle Paul teaches his nephews about mixtures, how to separate mixtures and combinations (i.e. reactions). Once the student finishes reading this chapter, discuss with him what he has read. You can ask questions like:

1. *How did the boys know that the powder was sulfur and that the filings were iron?*
2. *What is a mixture?*
3. *What were the two methods the boys used to separate the mixture?*
4. *What is a combination?*

Then have the student define mixture, solution, solvent and reaction and have him draw a diagram of one of the methods that the boys used to separate the mixture. Be sure that he labels all the key parts of his drawing.

Writing

You can have the student write a brief summary of what he reads in *The Wonder Book of Chemistry* or read several pages from an encyclopedia about the subject and then outline what he reads. For this week, have the student read about separation from the *Usborne Science Encyclopedia*, the *DK Eyewitness Chemistry* or the *Kingfisher Science Encyclopedia*. Then have him create an outline from that text. A typical seventh grade student could write:

I. *Separating a mixture breaks it into its components.*
   A. *The method you use will depend upon the properties of the substances in the mixture.*

II. *There are several different methods that you can use to separate a mixture.*
   A. *Decantation is used to separate insoluble solids from a liquid by letting the particles settle.*
   B. *Filtration uses a filter to trap insoluble solids and*

remove them from a liquid.
C. *Chromatography separates substances in a solution by the rate at which they move through a medium.*
D. *Evaporation is a method used to separate soluble solids from a liquid.*
E. *Distillation is a way of obtaining a pure solvent from a solution.*
F. *Centrifuging is a technique used to separate solids that are in suspension.*

Hands-On Inquiry

Laboratory Experiment: Separating Mixtures

You will need a bunch of small twigs and straight pins mixed together, a magnet and a glass of water. Explain to the student that he is going to perform the same separation tests that Uncle Paul and the boys did in the chapter he read. Then let him use the magnet to separate a portion of the pins. Once he is finished mix the pins and twigs back together and have him add half a cup of the mixture to a glass of water. The twigs should float to the surface, while the pins will sink to the bottom. After the student finishes, discuss his results with him and have him write an experiment report.

Around the Web

Have the student watch the following video series on "Separation Techniques used in Chemistry":

- **Chromatography Episode 1:** http://www.youtube.com/watch?v=1EIW_ZLJWWc

- **Chromatography Episode 2:** http://www.youtube.

com/watch?v=HqfEWb-OM58&feature=relmfu

There is no sound to these videos, but they show how applicable chromatography is to life. If the student is interested, have him repeat the experiment shown in the video using two different pens. Use each pen to make a dot at the bottom of a coffee filter. Then place the bottom edge into some rubbing alcohol, making sure that the dots don't touch the liquid. Watch to see what happens as the ink separates.

Quizzes or Tests

Give the student an oral quiz by asking him to define what a mixture is and to name four ways of separating mixtures.

The Nature Study Method

Comparable to the elementary years, the Nature Study approach to middle school science education is centered on finding science in nature and is enriched with reading from non-fiction books. The student will spend time outdoors studying the principles of science found in nature. Then, he will read about the subject in an encyclopedia and write about what he has learned. If time and interest allow, you can have the student research the topic around the web. What follows is a sample week using the Nature Study method of middle school science instruction for gravity.

Hands-On Inquiry

You will begin by taking a walk outside with the student, looking for objects in nature that are round and about the same size but different weights, such as a piece of fruit, a rock and a nut. Have him identify and collect at least two of these objects for later use. You can also have him identify and collect any leaves or flowers that he finds interesting.

Once you get home, have the student hold each of the round objects in his hand and drop them at the same time. What happened? (*He should see that both of the objects hit the ground at the very same time. If you can do this safely from a porch or balcony which will give you a bit more height, your results will be even more amazing.*)

Then spend some time discussing gravity with the student, you can bring up Galileo, who did the very same experiment he just did using one heavy ball and one light ball dropped from the top of the Leaning Tower of Pisa. If the student understands what you are discussing and you want to add more, have him drop a round object and one of the large leaves that he found. Then, discuss how air resistance affected his results (i.e. The wider an object, the more air resistance it will experience and therefore the more the effects of gravity will be altered.)

You can choose to have the student write an experiment report in his nature journal, or you can just have him write several sentences about what he has learned. A typical eighth grade student could write:

> *"I dropped an oak-tree acorn and a quartz rock of the same size from our deck. They both hit the ground at the same time, even though the rock weighed more than the acorn. Galileo did the same thing with two cannonballs of differing weights from the top of the Leaning Tower of Pisa. He is*

credited with discovering that gravity exists."

Information

Have the student read about gravity from a science encyclopedia, like *Usborne Science Encyclopedia, the Kingfisher Science Encyclopedia,* or the *DK Encyclopedia of Science.* Alternatively, you could have him read a biography on Galileo or Isaac Newton (Galileo discovered the concept of gravity, while Isaac Newton devised a mathematical way to calculate the force of gravity). When he is finished reading, discuss with the student what he learned about gravity.

Then have the student define gravity, weight and mass as well as draw a diagram of the forces at work in his nature study. His drawing should include a picture of the objects he was dropping along with arrows showing the forces of gravity and air resistance. You may want to also have him include the fact that the acceleration due to Earth's gravity is always equal to 32 feet per second (or 9.8 m/s).

Writing

When you use the Nature Study method, it is best to have the student use a blank journal for his work. The nature journal should allow for freedom to include what the student wants to incorporate. He can paste samples or pictures in there; he can also include his sketches or write about what he has learned. For this week, after he has completed the reading and discussion time with you, have the student outline the pages he read about gravity. A typical eighth grader could write:

I. All masses attract each other, which is called the force of gravity.
A. The strength of the force of gravity depends upon their

masses.

- *B. The force of gravity between two objects only becomes apparent when one or both of the objects have large mass.*

II. The force of gravity remains the same.

- *A. Different sized objects would fall at the same rate if only the force of gravity was acting upon them.*
- *B. Air resistance can slow objects down, reducing the acceleration caused by the force of gravity.*

III. Isaac Newton wrote the law of gravitation.

- *A. The law of gravitation connects the force of gravity between two objects with their masses and the distance between them.*
- *B. Newton's law allows us to predict the effect of gravity on objects.*

Around the Web

Have the student research gravity on the internet. Here's a few websites to get him started:

- **How Stuff Works on Gravity:** http://science.howstuffworks.com/environmental/earth/geophysics/question232.htm
- **Gravity-A Google Experiment:** http://mrdoob.com/projects/chromeexperiments/google_gravity/
- **Isaac Newton and the Law of Gravitation:** http://csep10.phys.utk.edu/astr161/lect/history/newtongrav.html

Quizzes or Tests

Give the student a matching quiz about gravity. You can

include the definitions of gravity, mass and weight and the value of acceleration on Earth due to gravity among the questions.

The Unit Study Method

The Unit Study approach to middle school science education is centered on a scientist whom the student explores through the basic components. The student will spend time reading about the scientist, he will write about him and complete a laboratory experiment or nature study related to the scientist he has been studying. If time and interest allow, you can have the student research the scientist on the Internet. What follows is a sample week using the Unit Study method of middle school science instruction for Galen.

Information

Have the student read a biography on the life of Galen such as *Galen and the Gateway to Medicine* by Jeanne Bendick. As he finishes each chapter you will need to discuss with him what he has read, which means that you need to also read the book. You want to ask comprehension questions and discussion questions that will make him think beyond the book. After reading chapter 1 of *Galen and the Gateway to Medicine*, your questions should be:

1. *Who was Galen?*
2. *How did he learn about medicine?*
3. *Why couldn't Galen see the inside of a human body to study the anatomy?*
4. *How did the ancient Greeks study science?*
5. *How does the ancient Greek method of studying science compare to our modern scientific method?*

Then, have the student define the field of medicine and physician. For the diagram portion of the information component, we recommend that you have him keep a detailed timeline of the scientist's life, adding whatever dates he can from each chapter.

Writing

Have the student take notes as he reads through the biography of the scientist. Then, once he finishes the book, have him look over his notes and create an outline for a brief report. Afterwards, the student should write a short paper on the scientist from his outline. His report should be anywhere from one to three pages long. A typical fifth grader could have written:

> *The title of the book I read is called 'Galen and the Gateway to Medicine'. It was written by Jeanne Bendick. He was considered the greatest physician during the Roman Empire.*
>
> *Galen figured out how the brain worked by using animals like monkeys, oxen, pigs, goats and many others. He also figured out how the heart worked. He treated four of the Roman Emperors. Without him the future would have been very different.*
>
> *Galen lived in Pergamum, which was in the mountains. His mother's name is unknown, but his father's name was Nicon. When he was in his 20's and 30's he became a doctor to gladiators. Pergamum was in the country now known as Turkey. There were very few good doctors there. Most doctors just announced that they were a doctor, and then just became one without any training. Galen actually had medical training.*

Galen learned to read and write Greek and Latin too. He also learned math and geometry. Greeks were wonderful at geometry because of the teachings of Archimedes, Euclid, Plato, Aristotle and many other famous mathematicians. Galen studied science also. He learned about the planets and the stars. Galen loved studying more than anything else. He had written three books by the time he was 13.

Galen did all of this work because he wanted to prove to the world that medicine and doctors are important. He wanted to be a great doctor. I believe Galen was a great doctor and scientist. He just got the lungs, stomach, and the heart in the wrong places in his drawings because animals are different from people. I think you should read 'Galen and the Gateway to Medicine' if you want to be either a doctor or a scientist.

Hands-On Inquiry

In the Unit Study method for the middle school years, you will typically be doing one major laboratory experiment or a multi-week nature study for each scientist. Galen pioneered the field of anatomy by performing less than humane dissections. He learned a lot about human anatomy through these dissections, but also got a few things wrong. For this study, have the student dissect a fetal pig. As he dissects the fetal pig, talk about the similarities to human anatomy and the differences. A good dissection kit will include this information for you to draw from. Once he completes the dissection have him write up an experiment report that includes a diagram of what he found inside the pig as well as the similarities and differences you discussed with him during the dissection.

Around the Web

Have the student research Galen on the Internet or in an encyclopedia. Have him add any additional information he learns about the scientist to his outline for his report. The following websites contain some good information about Galen...

- **Galen & Greek Medicine:** http://www.nlm.nih.gov/hmd/greek/greek_galen.html
- **Galen of Pergamum from Science World:** http://scienceworld.wolfram.com/biography/Galen.html
- **Galen from the University of Virginia:** http://www.hsl.virginia.edu/historical/artifacts/antiqua/galen.cfm

Quizzes or Tests

Give the student a test on Galen and his major discoveries, once he has completed the reading the book and writing his report.

The Science Project

In essence the Science Project is an in-depth experiment which will take several weeks to complete and follows the steps of the scientific method. Below is a detailed explanation of the process with a sample project from the field of astronomy.

Step 1

Have the student begin by choosing a topic in the field of astronomy that interests him, such as comets. He may need to narrow his topic down, such as, "I want to know what determines how fast a comet will melt and form a tail." Then, have the student think about a question he would like to know the answer to, such as, "*Does the size of a comet affect how fast it melts?*"

Step 2

Next, have the student do some research about his topic so that he can make an educated guess (hypothesis) on the answer to his question. For the question stated above, he would need to research comets and how they form their tails. Have him begin by looking the topic up in the references you have at home, and then make a trip to the library to learn more about the topic. The student should write any interesting facts he learns on index cards. Once he has finished his research, help him organize the index cards and write a brief one to three paragraph summary on what he has learned.

Step 3

At this point the student is ready to make a hypothesis or an educated guess about the answer to your question. A hypothesis for the question asked in step 1 could be, "*The more a comet weighs, the quicker it will melt*".

Step 4

Once the student has formulated his hypothesis, he needs to design his experiment. You will need to explain to him that his experiment must be a test that will find the answer to his question. It needs to have a control group as well as several test groups. The control group will have nothing changed, while each of the test groups will change only one factor at a time. An experiment to test the hypothesis given above would be to fill 3 different sizes of balloons (large, medium and small) with varying amounts of water. Make sure the student has 3 of each size to make the test valid before he freezes each one. Once they are frozen, apply a heat source for a set amount of time. Each time, the student will measure how much water has melted and record the amount. Step 4 may take several weeks depending on how long he designs his experiment to take. Be sure to have the student take pictures along the way and to record any observations he sees during his testing.

Step 5

As the experiment draws to a close, the student needs to compile his observations and results. Remember that observations are the record of the things you see happening in your experiment, while results are specific and measurable. His results can be recorded in chart form or he can just be a simple description of what happened in his experiment (*see Chapter 1 for an example of possible observations and results*). It is important to note that he is not interpreting the results of his experiment yet. He is simply relaying information on what happened during your experiment.

Step 6

Now that the student has his results, he can use them to answer his question. You also want to have him share what he has learned from the project in his conclusion. A possible brief conclusion to the comet experiment above would be, "*Comets melt at the same rate no matter what they weigh.*"

It is important to note that if the student was not able to answer the question he posed in step 1, he will need to go back and do some additional testing.

If the student has all the information he needs, have him begin to work on a presentation. He should prepare a brief 5 minute talk and a presentation board for his project. The talk should include the question he tried to answer, his hypothesis, a brief explanation of the experiment, the results, and the conclusion to his project. The board should be made out of sturdy material so that it will stand on a table on its own. It should include several pictures and graphs from the experiment and present the material in a visually appealing way, since it will serve as the visual aid for his presentation. After the student has completed his project board, help him to determine if he should include a part of the experiment in his presentation.

Once the student has finished preparing his talk and presentation, have him practice with you. Be sure to give him feedback so that he can make the necessary changes before he presents the project to a group.

Conclusion

The middle school years are not just mere place holders in a student's educational journey. They are critical for setting up the student for success in high school science. If you keep the goals we suggested in mind and base your curriculum on the components we laid out, you will lead the student to excellence in his science education.

5
The High School Years

The high school student has already mastered the basic facts as well as some of the reasons behind them, so he is prepared to move onto the more complicated details of science. This student is far more contemplative and when he asks questions, it is obvious that he has put some thought into the inquiry. However, the high school student is still working on knowing how and when to express himself.

The high school years begin when the student enters high school, usually around 9th grade, and end when the student is ready to go to college. During these years, the student is learning and applying the principles of science, as well as learning how to articulate himself intelligently. The student who has followed our plan up till know will have a good grasp of science, so he can really focus on tackling the difficult concepts along with shoring up his knowledge of the scientific method.

Your Goals

You have two goals for the high school years:

1. To make sure that the student knows and understands the principles and laws at work in science.
2. To teach him how to relate what he has learned to the things he sees around him.

We compare the high school pupil to a law student. He has access to a great deal of organized, filed away information, but he is still learning advanced techniques, as well as learning when to use the material and how to apply it. So, when you teach science, you will be playing to his intellectual strengths while teaching him how to filter what he knows and apply it to the current situation. You can also use science to work on the basic skills of research and note-taking at this stage.

The Components

The components you use during the high school years will be dependent upon what the student wants to do in the future. Let us be clear in saying that we believe that all students should take four years of science in high school. As we have stated before in this book, science is beneficial for training the brain to think logically while giving every student a healthy awareness of the world around him. However, by the time a student reaches the high school years, he is starting to get a clearer picture of what he would like to do. If the student you are teaching is scientifically inclined, his curriculum should contain all of the basic and optional components suggested below. If he is not, the basic components should be more than enough to meet your goals for these years.

With that said, science during the high school years has three basic components and two optional components. The basic components are:

1. Textbook with Experiments
2. Events in Science
3. Exams

The optional components are:

1. In-depth Project
2. Research Report

How you put these components together will vary on the method you choose, which we will share more about later in the chapter; for now, let's look closer at each of the components.

The Basic Components

The three basic components are a textbook with experiments, events in science and exams. A good science curriculum for the high school years must have each of the following components.

Textbook with Experiments

The high school student needs to study science from a standard high school textbook, such as Campbell's, Prentice Hall or Science Matters. He will complete one text, or discipline, per year by reading through a section each week. If you want the student to work independently, have him take notes as he reads through each portion. Then you can look over his notes or have him write a brief report about what he has learned once he is finished the full chapter.

Alternatively, you can give a lecture for each section that will highlight the important concepts as well as introduce other related information. The student should definitely take notes, but we recommend that you provide him with a skeleton of your lecture, if he is relatively new to note-taking. You can choose to assign the text reading before or after the lecture.

The student will also need to do experiments that coordinate with what he is reading about in the textbook. Some of the texts mentioned above will already have experiments suggested in them, but for others, you will need to acquire experiments from another source or create your own. As a part of each experiment we recommend that the student completes a lab report. These reports will be a more sophisticated version of the experiment reports he wrote during the middle school years. We also recommend at least 2 to 4 times per year, the student writes a full lab report, including a section on research (*see the Appendix for more information on lab reports*).

Events in Science

It is important that the high school student becomes familiar with the people and events that have shaped the disciplines of science. This will give him a deeper appreciation for how far the subject has come as well as teach him how it relates to his everyday life. There are several ways you could approach this component, some more rigorous than others. For a more demanding curriculum, have the student read several articles in current scientific journals that are related to what he is studying. For a more relaxed approach, have the student read books that have been written about the great scientists or the major events that have occurred in a given field. You can also have him follow the news and summarize any of the topics that relate to science.

Exams

If you have noticed so far in this book, we have not included exams as a basic component of science education. The main purpose of the elementary and middle school years is to build the student's scientific aptitude and confidence, so we believe that tests up until this point should be an optional component which the teacher determines whether or not to employ. The student is doing enough work during these years that you can use to grade or gauge his progress.

However, the high school student is becoming more and more independent with his work, so testing is necessary to affirm that he has learned the concepts. Plus, he will be expected to know how to perform on a test as he progresses towards the college level. So, for the high school years, we have included exams as part of the basic components. The questions on the exams should be mathematical problems, short answer and essay questions. They should require the student to think and draw upon the knowledge he has gained.

The Optional Components

The student who desires to major in a science-related field should have a curriculum that includes at least one, if not both, of these components. These will prepare the student for the rigors of college-level science as well as give him a strong foundation for his future career.

In-depth Project

The in-depth project will have all the components of the scientific method, but will be more complex than the science project we discussed for the middle school years. The student will still come up with a question, do some research, make a hypothesis, design an experiment, analyze his results and draw a conclusion, but on a much deeper level. In the middle school years, the student might have spent 4-6 weeks on his science project, but during the high school years, his in-depth project should take about a semester to complete. He will be looking at least 2-3 variables and it will take several experiments to fully test his hypothesis. The in-depth project will look a lot like Gregor Mendel's study of peas, where he sought find out if there was a relationship between generations, ultimately discovering the field of genetics. We are not saying that the in-depth project needs to lead to a Nobel Peace Prize winning discovery, just that it should take time and several tries to reach the answer to the student's question.

Research Report

Every high school student that is interested in further science study should be doing at least one research report per year in the area of science. The reports should take anywhere from 6 weeks to several months to complete and should be loosely associated with what he is learning for the year. In other words, if he is studying biology, his topic should be related to the field of biology. The student can choose to research a major concept, a scientist or an important discovery within the field he is studying. The research report must include a thesis statement which causes the student to form and defend an opinion about the material. His completed paper should be 6 to 8 pages in length. It should touch on why he chose the topic and how it affects him as well as thoroughly explain what he has found out about the subject.

A Word about Math and Science

In order to excel in science, the high school student will need a strong foundation in math. Many of the calculations in physics and chemistry require that the student has at least a working knowledge of algebra. In addition to algebra, the student who desires to go further in the sciences should aim to complete the first year of calculus in high school as many of the more advanced formulas require a working knowledge of the subject.

What the Student Needs to Learn During the High School Years

The high school student is working on learning the more complex information found in a scientific discipline as well as learning how and when to apply his knowledge. Here's a list by discipline of key facts that he should know before leaving the high school years. *(Note: This list contains what all students must know, the science-oriented student will need to go a bit deeper than what is on this list.)*

Biology

- know everything from the elementary and middle schools lists
- know how the various living things regulate the functions of their body
- be familiar with the systems of the human body and how they function
- understand biological structure (i.e. cells form tissues,

tissues form organs and so on)

- know what each of the components of a cell do
- know how cells reproduce and divide (mitosis and meiosis)
- know what DNA and RNA are, how they reproduce and what they are responsible for
- know what chromosomes are and what role they play in genetics
- be familiar with how genetics works (i.e. dominant and recessive genes, the Punnett square and so on)
- know what biological adaptations and biodiversity are as they relate to ecosystems
- be familiar with what it means to conserve and manage our biological resources
- be familiar with the theory of evolution and be aware of the theory of intelligent design
- be familiar with the concept of natural selection

Earth Science

- know everything from the elementary and middle schools lists
- be comfortable with reading and drawing maps
- know about plate tectonics and the movement of the Earth
- be familiar with convection cycles

- be familiar with hydrology (i.e. how streams, ground water and so on function)
- know Earth's natural resources, how they are obtained and how we use them
- know about fossils, how they are formed and radioactive dating
- be familiar with weather variables (i.e. El Niño and so on) and factors that can affect climate

Astronomy

- know everything from the elementary and middle schools lists
- be familiar with the major events of astronomy (i.e. the Copernican Revolution and so on)
- know with the major characteristics of each of the components of the universe
- know the major characteristics of each of the components of our Solar System
- be familiar with the Big Bang Theory
- be comfortable with using a telescope to identify the celestial bodies in the night sky
- know the phases of the moon and how they affect tides
- know about the arc of the Sun (i.e. how it travels in the sky) as well as how this affects the Earth

Chemistry

- know everything from the elementary and middle schools lists
- be familiar with mixtures, solubility, solvents and solutes
- know the subatomic particles and how an atom is structured
- be familiar with the various atomic models that have been proposed over the years (i.e. the Thomson model, the Rutherford model, the Bohr model and so on)
- know how and why atoms bond as well as about molecular attraction
- know about the Law of Conservation of Matter
- be familiar with chemical formulas, chemical equations, moles and stoichiometry
- know about chemical equilibrium and reaction rates
- be familiar with trends found in the periodic table
- be familiar with radioactivity and nuclear energy
- know about thermodynamics
- know about electrolytes and reactions that involve acids and bases
- be familiar with redox and how to balance an equation
- be familiar with carbon chemistry and be able and identify the formula, name and structure of several

common organic compounds, such as methane, carbon dioxide and so on

Physics

- know everything from the elementary and middle schools lists
- know the various units of measurement and be familiar with converting them back and forth
- know how to evaluate and graph scientific data
- know the mathematics of motion (i.e. calculating speed, velocity, acceleration and so on)
- be familiar with kinematics, statics and dynamics
- be familiar with the different types of motion (i.e. two-dimensional, circular and so on)
- know the Law of Conservation of Energy and Momentum and Hooke's Law
- know about work, energy and power and the mathematics behind the various types of energy
- know Coulomb's Law and Ohm's Law as well as how electric fields and currents work
- know the various parts of an electrical circuit as well as how they can be put together (i.e. in series or in parallel)
- be familiar with magnetic fields
- know the types of waves and wave phenomenon

- know the speed of light, the index of refraction and Snell's Law
- be familiar with quantum theory

What Can It Look Like

During the high school years, your options for teaching science are going to be more focused than they were during the elementary and middle school years. The method you choose will be solely based on what the student plans to do in the future. The two methods for the high school years are the Knowledge Building method and the Science Oriented method, which are both explained in greater detail below. As before, we have included a sample week broken down into the components we explained above for each method so that you can see what each routine would look like. (*Note: If the student is undecided, we would recommend that you go with the Science Oriented method.*)

The Knowledge Building Method

The Knowledge Building method for high school science focuses on constructing the student's understanding of the subject. It uses textbooks with experiments to teach him about the principles and laws at work in science, events in science to teach him how science relates to the world around him and exams to test what he has learned. What follows is a sample unit for using the Knowledge Building method of science instruction for high school. The following work should be completed over one and half to two weeks.

Textbook with Experiments

Have the student read chapter 2 on "Kinematics" in

Physics Matters by Marshall Cavendish. He will be reading about speed, velocity and acceleration as well as the speed/time graph and the acceleration of a free fall. If you choose to give a lecture on the subject, make sure you touch on Newton's 3 laws of Motion, the differences between speed, velocity and acceleration, how to create a speed/time graph and how these concepts affect the student. Have him take notes from his reading and from the lecture, if you give one. Once he has completed the reading, have him complete the chapter worksheets in the text and go over the answers with you. After that, have the student do experiment #5 from the Practical Book and answer the questions associated with the lab in the book.

Events in Science

Have the student find several short biographies on Isaac Newton on the Internet or at the library. Have him read and take notes on what he has found. Then, have the student write a 1-2 page report on Isaac Newton's contributions to science.

Exams

Have the student complete the associated worksheets from the *Physics Matters* workbook for a grade. There are three worksheets for this chapter, so you could assign two for practice and one as a test, or create a test of your own for the student to complete.

The Science Oriented Method

The Science Oriented method for high school focuses on building the student's understanding of science as well as feeding his passion for the subject. It still uses textbooks with experiments to teach him about the principles and laws at work in science, events in science to teach him how science relates to the world around him and exams to test what he has learned. However, the student following the Science Oriented method will also add in an in-depth project and a research report to fill his desire to learn about the subject. What follows is a sample week for using the Science Oriented method of science instruction for high school.

Textbook with Experiments

Have the student read Chapter 16 in Campbell's *Biology: Concepts & Connections*. There is a lot of information in the chapter, so you may want to have him skip certain sections. You will need to give a lecture pulling out the key ideas on prokaryotes and protists that you want the student to know. After that have the student look at microscope slides of bacteria, algae and amoeba. He should also complete a brief lab report explaining what he did along with drawings of what he observed in the microscope.

Events in Science

Have the student research how algae can produce biodiesel fuel. He can use the internet or the library to find periodicals that contain recent articles on the subject. Then, have the student write a 1-2 page report about his findings.

Exams

Have the student complete the chapter review as a test or write your own exam for him.

In-depth Project

For the in-depth project, the student will follow the same steps we laid out in chapter 4 for the science project.

However, during the high school years he will go deeper with each step. For example, in middle school he may have had only 5 references, in high school he will need around 15-20 references for his research. The in-depth project should also be a semester long process, rather than several weeks.

So, let's say the student chooses to do an in-depth project involving hydroponics. He wants to know if the use of hydroponics will increase plant yield. His hypothesis, which is supported by his research, states that "*Hydroponics produces a superior yield to traditional growing methods.*" For his experiment, he sets up two growing environments, one in potting soil and one using hydroponics. At the middle school level the student would have only used one type of plant observed over two to three weeks. Now, for the in-depth project, he will use multiple types of plants, i.e. a flower, grass, lettuce and a vegetable, which he allows to grow for two to three months. The student will still need to record his observations and data daily.

Once his testing is completed, the student will use his mathematical knowledge to analyze and report his findings. His conclusions will be much more in-depth and will include his own inferences about the findings. Once his project is complete, the student should give a 10 to 15 minute oral presentation explaining his project and his results.

Research Report

The research report process should take the student about half of the year to complete. Begin by having him pick a topic and research it, finding out as much material as he can. He can look in biological abstracts, Google Scholar, reference books and encyclopedias. As he finds information, have him take notes that are separated into subtopics. We recommend that he put the different pieces of information onto index cards that are numbered for each reference and subtopic. However, if you want the student to use a computer program rather than hand–written notes, we recommend RefWorks, which is widely used as a reference software in colleges today.

The next step is to have the student write his thesis statement. The purpose of the thesis statement is to give a focus to his paper. His statement should give his point of view or slant on the topic. You can ask him the following questions to help him craft a thesis statement:

- *What do you know currently about the topic?*
- *What are the questions that you have about the topic?*
- *How do you feel about the topic?*

This is a fluid process, so his thesis statement may need to be revised several times before the first draft is written.

After the student has written his thesis statement, he needs to create an outline for his paper from the information that he gathered. His paper needs to have three sections:

1. **Introduction:** This section gives a brief look at the topic, states his thesis statement and explains why he chose the topic.
2. **Body:** This is the main part of his paper which contains multiple paragraphs full of information that support his thesis statement. The body should include several quotes from experts or excerpts from his research that give credence to his thesis statement.

3. **Conclusion:** This section will restate his thesis statement, summarize his supporting information and apply it to today.

After you have approved his outline, have the student turn in a rough draft of the paper. If he is not familiar with writing research papers, you may want to have him turn in multiple drafts. Either way, the final research report should be 6-8 pages in length (double-spaced). You are looking to make sure that the paper is written in the third person, that it uses the correct MLA style documentation, and that the paper has a strong thesis statement with good supporting evidence.

Conclusion

The high school years are the culmination of the student's foundational years of his academic journey. They are essential for teaching the student to apply the principles of science and to be able to analyze the data he is receiving. If you keep the goals we suggested in mind and base your curriculum on the components we laid out, you will lead the student to excellence in his science education.

Part 3: A Look at Current Science Education Methods

6
The State of Science Education

Currently North America is struggling with teaching science to our young people and our country is at an impasse about what to do about it. Looking at the statistics that we see in the elementary years, U.S. students perform in the top 3 percentage of standardized tests. As they get to middle school that drops to the top 10, and as they move to high school it drops again to the top 25. We as a nation are struggling with how to teach science to our young people.

We see some startling shortfalls in science education when we look at a study that was done by the Center for the Study of Mathematics Curriculum in 2005. The study found that the curriculum is becoming homogenized and diluted. They also found that college prep courses, although available, are not being taken by the majority of students, and that time spent on homework has declined. The researchers also pointed out that...

> *"...compared to other nations, American children spend much less time on schoolwork; time spent in the classroom and on homework is often used ineffectively; and schools are not doing enough to*

help students develop either the study skills required to use time well or the willingness to spend more time on school work."

After reading this we should all be concerned about our children's future as it pertains to science education. If we want to provide a superior education to our children and increase our standing as a nation in science, we must make a change.

Our lack of excellence in science education does not only affect our standing in the world as a leader in innovation but also has wide reaching effects concerning our nation's security and our edge in business. Let's look at some statistics from Forbes.com and the National Research Council:

- In 2009, for the first time, over half of U.S. patents were awarded to non-U.S. [based] companies. *(Danger: America Is Losing Its Edge In Innovation, Norm Augustine, 2011)*
- China has replaced the U.S. as the world's number one high-technology exporter. *(Danger: America Is Losing Its Edge In Innovation, Norm Augustine, 2011)*
- Between 1996 and 1999, 157 new drugs were approved in the U.S. Ten years later, that number had dropped to 74. *(Danger: America Is Losing Its Edge In Innovation, Norm Augustine, 2011)*
- Nearly 70% of engineers with a PhD [from U.S. universities] are foreign-born. *(Rising Above the Gathering Storm The National Academies Press, 2010.)*
- 52% of Silicon Valley's start-ups during the recent tech boom are [from] foreign born [individuals]. *(Rising*

Above the Gathering Storm The National Academies Press, 2010.)

- Only 6% of Indian, 10% of Chinese, and 15% of European students graduating in 2008 [from U.S. universities] planned to settle in the U.S. *(Rising Above the Gathering Storm The National Academies Press, 2010.)*

These statistics should concern you. Students from other parts of the world are learning science at a more advanced level than here in our country, which leads one to conclude that there must be some gaps or problems in our system of science education. We must find a solution to this and address it before we fall further behind. In our ever-changing culture where science and math are crucial to understanding our world and to the success of our future, this is not acceptable. We have to strive to be the very best in science education in order to maintain our edge as a global leader in innovation.

The Challenge

We believe that one of the biggest challenges which students face today is the overwhelming information that is available to them. From TV to the Internet to social media, children today must assimilate and process vast quantities of information that in the past were only contained in books and journals. Most students today do not know how to filter the information that they receive each and every day.

Consider the analogy of a filing cabinet full of millions of unorganized papers, some that are junk, some that are absolutely necessary. If you were looking for a single piece of paper, it would take quite a long time. However, if you had a system in place that would sort each new piece of paper that came in as junk or necessary, and then the system further sorted the necessary papers into categories, it would be far easier to find a single piece when you were looking for it. Our students are being assaulted with thousands of pieces of information each day and unless we teach them how to organize this information, it will simply become a huge pile to wade through every time they are trying to recall a fact. We must do a better job of teaching our students a system that will help them to filter the information that they are receiving if we hope for them to be successful in the 21st century.

Current Successes

Let's examine the educational systems and strategies employed by some of the successful schools found in other countries. It is widely accepted that Asian schools, especially those found in Japan, South Korea, China, Hong Kong, Taiwan and Singapore, get exceptional performance from their students because of their discipline methods and rigorous preparations. Asian students typically devote a great deal of their time to studying and memorizing facts and formulas in order to produce the results that are expected from them, which has led to Asian students being ranked among the highest in the world. As adults, they go on to be some of the best and most sought-after science professionals in global industries.

However, applying this kind of strategy in the American setting might not work. We are a nation that has always taken pride in the fact that our children's education is an enjoyable yet useful activity for them. We also pride ourselves in being a country of independent thinkers that are known throughout the world for their complex problem solving. Many Asian students are coming to the United States for university because they have found that all the studying and memorizing has not prepared them to think outside the box and or given them complex problems solving skills. So we see that the Asian educational philosophy has downsides as well.

Instead of looking to Asia for the answers, how about looking for a Western country that is producing successful students? Over the last decade, Finland has been recognized to have a comparatively successful school system. However, their methods are very different from the traditional "no pain, no gain" belief of many schools, including most of the educational institutions operating in our country.

Few educators in the U.S. are familiar with the unorthodox methods employed by the Finnish educational system, but these methods have propelled their students to the top of the world ranks. Lynelle Hancock says in her article entitled "Why are Finland's Schools so Successful?" that Finland has "vastly improved in reading, math, and science literacy over the past decade in large part because its teachers are trusted to do whatever it takes to turn young lives around." Education is a two-way process; both the students and teachers have to be involved to ensure that there will be reliable results. However, in our current educational system bureaucracy and "tried-and-tested" teaching methods do not allow much room for bright young teachers to experiment with how they want to teach very important yet sometimes difficult subjects.

In today's fast-changing environment, students cannot be expected to take instructions and lessons like we used to do. Young people are becoming more and more individualized in their tastes and attitudes toward the world and the people that surround them. This means that teachers have to reconsider their teaching strategies and methods in order to adapt to the changing world of their students. This willingness to adapt is Finland's secret to academic success.

The Finnish Model

Finland has one of the highest-performing school systems in the world, as measured by the Programme for International Student Assessment (PISA), which assesses the reading, mathematical literacy and scientific literacy of fifteen-year-old students in the thirty-four nations of the Organization for Economic Cooperation and Development (OECD), which includes the United States. In *Finnish Lessons: What Can the World Learn from Educational Change in Finland?*, Pasi Sahlberg explains how his nation's schools became successful. As a government official, researcher, and former mathematics and science teacher, Sahlberg attributes the improvement of Finnish schools to bold decisions made in the 1960's and 1970's.

Here are a few key features of the Finnish educational systems that we have gleaned from Sahlberg's book...

- *Finnish students do not take home assignments or take exams until they are teenagers.* This strategy is perhaps responsible for ingraining in the students' minds and psyche that learning is a fun and worry-free activity that they can enjoy with their peers and their teachers. By taking away the pressure to perform well in school and compete with each other academically, the students feel comfortable with the notion of studying for their own personal growth and not because they need to fulfill the demanding expectations of their elders.

- *School children are not measured for their academic performance until they have finished six years of education.* Grades can be a source of intense pressure for younger students, especially when they are not getting the grades that they are "supposed" to get. Getting lower grades might dampen their confidence and enthusiasm for school and can lead them to believe that they are "slow" or "stupid". Children learn at different paces, and this is important for educators and parents to understand because a structured grading system can be discouraging to students instead of encouraging.

- *A huge chunk of Finnish students get extra help with their schoolwork during the first nine years of their education.* This can make all the difference in the world to many children as they need a solid foundation to base their entire academic futures upon. Helping them form a solid educational foundation is the first step toward developing curiosity and enthusiasm in "difficult" subjects such as math and science.

- *Science classes are limited to 16 students per class.* This is perhaps the easiest strategy that we can derive from the Finnish school system. With fewer students in science classes, the teachers can focus on designing and structuring practical experiments that the students can perform in order to learn various principles of science. Allowing the students to "discover" something for themselves instead of dictating sets and sets of facts for them to memorize will ensure that the subsequent knowledge will stay with the student for years to come.

- *The Finnish school system is funded entirely by the*

government. This just shows how much money and effort the Finnish government is devoting to their educational system, which probably contributes to the improvement of the schools and the confidence of the citizens with their educational institutions.

- *The curriculum guidelines set by the government are broad instead of definitive rules.* This allows the teachers to be creative and resourceful in the planning of their lessons and teaching materials. This gives them the space and opportunity to design more creative, fun, and "edutaining" activities that will enhance and improve the students' learning experiences.

- *Finnish teachers are also given the same status as doctors and lawyers.* Confidence and pride in what they do is very important for teaching professionals because it leads to them becoming more enthusiastic in their work and in their mentor-mentee relationship with the students.

- *Many Finnish schools are small enough that the teachers know every student.* The personalized approach to learning benefits the student the most, because he or she will not be intimidated to ask questions or raise issues. The teacher will also be able to supervise the progress of the each and every student.

Although not every method used in the Finnish educational system will work in the American context, by examining their structure, we can get a good idea about what might work be beneficial to our own students.

Conclusion

While we cannot adapt every successful Finnish or Asian method in our local schools, we can reexamine our current educational system to find out which areas are hindering our student's ability to learn and then make adjustments in those areas. In this ever-changing world, the best thing educators and parents can do for our students is to make sure that the educational system is evolving with their skills and needs to help them compete with the rest of the world. Learning should not always be easy, but we can try to make it more fun and interesting for the students, which will cause them to be more involved and enthusiastic about the learning process.

7

Current Challenges in Science Education

Science should be one of the most basic subjects that students learn in school, aside from reading and math. It is the same in just about every country around the world; though the language used to teach and learn science might be different per region or area, the basics remain the same. Even with the uniformity of the subject, different countries and educational systems around the world teach science in a variety of ways. We know that there is always room for improvement in the methods of instruction used by our educational system, and this is especially true for the field of science education.

The Problems with the American Education System in Regards to Science Instruction

In an article for *The Atlantic*, the former Chancellor of the New York City Department of Education Joel Klein calls out American schools for their "failures" in educating our youth. "There are very real risks," he says, "risks that can eventually affect the future of the country—involved with a system of education that cannot cope with the evolving minds of the students." In *A Nation at Risk*, the report filed by the National Commission on Excellence in Education thirty years ago, education experts warned that there is a "rising tide of mediocrity that threatens our very future as a nation and a people." Surprisingly, there has been very little progress since then, even though the government has doubled the amount of funds that go into K to 12 public education.

In the National Assessment of Educational Progress, only a small chunk of eighth-grade students were competent in math, science, and reading. Although there are a good majority of high school students that have the opportunity to graduate, we are still far behind the other industrialized countries who can boast of nearly perfect percentages in science scores.

Currently, the country is at an impasse; we have been using antiquated methods that do not produce results. These methods are barely enough for sustaining whatever we have in the educational system now. American schoolchildren are not performing as well as their counterparts in Europe and Asia. This is very alarming because there are a lot of funds being funneled into their education, but very little to show for it. If the funds are not lacking, it is only logical to assume that the bulk of the problem rests in the methods that we employ to teach.

The world around us is rapidly changing. Holding on to "tried and tested" methods is not the right step. However, in the face of chaos and confusion, the natural tendency is to hold on to tradition. Despite all the funds that are being funneled into the purpose of providing education for our children, teachers still remain some of the lowest-paid professionals in the country. On top of that, they are evaluated not for the creativity, resourcefulness, and overall impact that they bring to the learning and intellectual development of the students, but rather for their student's test scores. As an indirect result, many young people are discouraged from teaching at the basic education level.

If a young college graduate aspiring to share his knowledge to the next generation wants to earn a decent living, he knows that he will fare better in the collegiate and university levels, where he will be given more than twice the salary that he expects to receive as an elementary or high school teacher. Other countries have remedied this by increasing the compensation and benefits to their teachers, allowing the educators to have almost the same professional reputation and credibility as doctors and lawyers.

But the problems with science education do not end with just increasing the amount of compensation that the educators get. There are a lot of factors involved. The first and most obvious problem with science-illiterate students is that they will not be able to compete with their counterparts from other countries in the future. Even though alternative methods of teaching (such as the ones developed and implemented by Finnish educators) do not emphasize competition as a main reason for educating students, when we compare our students' aptitude to the aptitude of students in other countries from similar age groups and school levels, we find the comparison to be quite disappointing.

Jean-Lou Chameau, the president of the California

Institute of Technology, offered his "theory" of why science education is poorly done in the United States. He says that the science teachers in this country do not "tend" to have science backgrounds; instead, they are often trained in general education with an emphasis in the sciences. Many science teachers in the K to 12 levels do not know the science subjects that they are teaching intimately, and this results in poor teaching of the subject. A teacher that is not well-conversed or well-trained in the subject that he is teaching cannot excel in relating his knowledge to his students.

Dr. Chameau's theory was posed on a radio show in 2009, which prompted many people to call the program in order to relate their own experiences as practicing science teachers. While a couple of these callers disagreed with Chameau's theory, some educators shared additional factors that are complicating science education in the country. As narrated by Chris Mooney in the article entitled "What Wrong with U.S. Science Education?" for the *Science Progress* website, some callers were said to have...

> *"...pointed out that the necessity of 'teaching to the test' often constrains the ability of science teachers to more creatively engage students. Similarly, others observed that many students are afraid of science and math, fearing it's too hard, and simply not for them."*

Insisting that people can be categorized as either a "science person" or a "math person" can also be a problem, because such statements imply that one is either born with the capability for understanding scientific concepts and theories or not. This causes the student to think that once they have done poorly in a particular science subject, they are "doomed" to do equally as poorly in all scientific disciplines. This kind of unfounded belief is what we want to dispel. All kinds of students can be helped to understand, if not excel in science subjects.

The problems of the country regarding science education cannot be solved by providing piecemeal or "band-aid" solutions that will temporarily alleviate the symptoms. There should be strong desire from the government, educational, and private sectors that will push for cultural change. Chris Mooney says in his book *Unscientific America: How Scientific Illiteracy Threatens Our Future* that:

> *"...at the core of that change must be the recognition that science doesn't have to be something weird, different, and alienating. It isn't just brainless memorization, and it isn't useless stuff that you'll never need."*

Many students in this country are not aware how fun and relevant science can be.

Threats of Scientific Illiteracy

Chris Mooney also states in his book *Unscientific America: How Scientific Illiteracy Threatens Our Future,* which he coauthored with Sheril Kirshenbaum, that:

> *"There's almost a kind of trap when it comes to teaching an intricate topic such as science. If you lose non-scientists in the weeds of information, they'll never see why it matters. But scientists thrive in the weeds—that's their job. Our science teachers, then,*

> *are a critical conduit between the two groups. They may or may not have scientific backgrounds, but if they can't trim the garden, they are bound to fail."*

There are several problems involved in not teaching our students well when it comes to their science subjects. These problems are not only with grades or world ranking, which are only used to gauge the aptitude of the students, but they are more subtle and serious than many of us would like to admit.

Students who are not encouraged to learn science, or have struggled with science due to problematic instruction, tend to be less curious than their counterparts. The scientific instinct to question why some things are this or that way is very important in the progress of not only academia but the rest of the country as well. Fewer students becoming interested in the sciences means that there will be less individuals who will be working in the scientific industry, thus seriously hindering the technological, medical and intellectual progress of the country. With fewer people being interested in knowing why, there will be a diminished push for new discoveries.

Science is also the direct enemy of ignorance. With a better understanding of science, we gain a better understanding of the world around us, which allows us to determine our place not only in the "natural order of things" but also in history. Science has always been instrumental in improving human lives ever since our species has existed. If not for the discovery of tools, the capacity to harness fire, the knowledge to farm and hunt or develop industry, we would all still be living in caves today. Science has made a lot of things possible and without the virtue of scientific curiosity, our species might have been long extinct.

A better understanding of science can also lead to a better understanding of other subjects as well. Almost every human profession has been rooted in science one way or the other. Science is a basic block upon which we can build a whole castle of human knowledge. Science is concerned with how the world works, as well as the major industries that make the planet more livable for us. Not understanding science correctly because it is not taught properly to students in school will be problematic for our society.

Conclusion

The main currency in science is people, not machinery, electronic devices or digital gadgets. It is the people, intelligent people that understand the basics of the scientific method and possess scientific curiosity, that truly allows science to grow, which in turn enables the world to develop further in knowledge. Even after thousands of years of science, we only know a small fraction of the knowledge that is available to us.

So how can we ensure that our students gain scientific literacy they need to succeed? Teachers and scientists can actively participate together in science education. The presence of scientists with real world experience of the principles that the teacher is seeking to relay can only increase the student's knowledge base. This partnership should happen often in our schools and both parties should be rewarded for their engagement.

In the higher levels of education, scientific literacy can also be promoted. There are a handful of colleges and universities requiring that all their college-level students should have lab-based science courses. This refreshes the scientific knowledge that the students have stored within their minds and promotes scientifically literate graduates.

We can also harness various kinds of technology that we have at our disposal today to improve scientific literacy. Students tend to be more interested in a slide show with moving elements than in staring at their teacher's face while he or she is talking about something that sounds so difficult. There are many programs and applications designed and developed for an educator's use, especially within the classroom where students spend much of their time. By doing so, we are increasing the chances for our students to use their scientific education as a basis of a more meaningful life and eventually, a better world in the future.

In the next chapter we will look at a new way to teach science that will address the current problems that we face.

8
How to Teach Science

In order to continue our country's innovation and help our students excel we must improve upon the way we teach science, but how can we do this?

In chapter 6 we used the analogy of an overflowing filing cabinet to show how much information must be processed by the young people of this generation. They are overwhelmed by the amount of data that is presented every day, from the TV, the Internet and from what they learn in school. All of this information can lead to an inability to process and filter important information from unimportant information. We must teach our children how to effectively screen incoming information and science can be one of the tools that will help our students learn how to filter the data they are constantly receiving. So, how should we go about teaching science?

A New Way to Teach Science

We must not focus solely on using only textbooks and references. We must also include science experiments along hands-on, tactile learning environments. Science is about exploring the world around us and learning how it operates. It is impractical and difficult to effectively learn about something like science without experiencing the subject you are studying.

For example, you would not teach someone how to jump rope by giving them only directions, and then expect them after only studying these directions from a book to go to a jumping rope competition and come in first place. Assuming that they have never seen a jump rope in action before, they would utterly fail in the competition because they have no understanding of the basic concepts of jumping rope. That would be unfair and impractical to the teacher as well as the student.

We should allow the students to "get their hands dirty" when learning about science by using inquiry-based methods. The National Science Education Standards say that...

> *"Scientific inquiry refers to the diverse ways in which scientists study the natural world and propose explanations based on the evidence derived from their work. Inquiry also refers to the activities of students in which they develop knowledge and understanding of scientific ideas, as well as an understanding of how scientists study the natural world."*

Research has shown that the student will learn a great deal more if they have real hands-on experience manipulating the world around them. According to a research study completed by Kimberly J. Vogt at the University of Dayton, children's test scores increased in science as hands-on activities were included into the lesson. Many of the students also said that their interest increased in the subject as well.

In a study done by Bredderman in 1983 which compared typical textbook science teaching to 3 types of hands-on science instruction, he found that students using the hands-on instruction method had higher overall test scores than the scores of the students using just textbooks to study science. Many of the common features of the alternative curricula used during the Bredderman study focused on more experimental, visual and tactile science as well as on the process of science rather than the typical textbook method.

Another study completed by Allen Ruby of the Rand Corporation in 2001 looked at the effectiveness of hands-on science curriculum. His analysis showed that there were significant positive differences between the scores of students learning through hands-on science versus textbook learning only. An interesting side note to this study by Allen Ruby is that there were no significant differences between student aptitudes. In other words, no matter the student's ability, using hands-on methods to teach science is still preferable.

We can see from these studies how important hands-on, process-focused science is to the future success of our students. However, as with all disciplines there will always be a need for basic content knowledge teaching (i.e. facts, theories and laws of science). Science cannot be taught without it but we must learn to combine the two approaches and develop a proper balance between content and experience.

Content instruction is often detailed and complex, but by using experiments and hands-on scientific demonstrations we can grow students' understanding and love of science. These practical demonstrations and manipulations will also help the student to see the purpose of learning some of the more complex and abstract laws of science.

Another important thing we need to consider is not to put all the pressure of teaching the students on the teachers within the classrooms. The educational system in our country could consider the "borrowing" of science professors and researchers from universities. These experts could engage the current science teachers in discussions and dialogues about scientific issues and help them to develop theoretical knowledge into practical experiments that could be safely performed inside the classroom. These scientists could contribute to the improvement of science education at the K to 12 level.

"Improving Science Education: The Role of Scientists" by Rodger W. Bybee and Cherilynn A. Morrow, says that scientists can be "involved in a variety of ways that accommodate their talents, time and interests and in ways that are ultimately helpful to the educational system." The paper mentions examples of when scientists have visited several different schools. These individuals taught a single lesson for each visit. However, these lessons were apparently not enough because after the initial fascination of the students, the lesson quickly faded from their mind. So, how do we make sure that the information is retained by the students? This seems to be the main problem of not only teaching science, but of many of the other subjects as well.

The Bybee and Morrow paper also states that the visits can be "helpful", but the authors still believe that there are "much broader and deeper ways that the expertise of scientists and engineers may contribute to educational reform." However, reform of this type is typically easier to apply at the college level, as webbing among faculty members is more possible and there are fewer rules and regulations that may or may not hinder the learning development of the students. The paper states: "(A)t the college level, faculty should consider developing collaborations with faculty in their school of education. Developing a real understanding of the issues faced in both disciplines can lead to better education for all students including future teachers." We agree that one of the best ways to incorporate scientific experts is through professional training.

One of the authors, Ms. Morrow, also proposed a framework in which scientists can act as a resource person for the basic educational system. She states that this is a "good intermediate level of involvement" that can definitely improve the teaching of science in the foundations of our educational system. Her framework provides the students with people and an environment that can help them develop a genuine interest in science. Without genuine interest, many students just go through their entire academic careers memorizing fact after fact and then completely deleting the information from their brains after they are done with school, or even worse, right after finishing the course.

Morrow's framework proposes that scientists and researchers act as judges in science fairs, where their expertise will be valued not only by the teachers but also by the students who will feel pride over having shown their projects to real scientists. Offering one's email address to the students in the case of questions can also be very useful, especially if the students will send in questions that are relevant to the expertise of the scientist. Giving a tour of a real research facility for students on a field trip can also help, since the students will be able to find out how "real" science works. This will give students a broader and "out of the textbook" glimpse at science that can help spark interest in their minds.

Teachers should also consider "scaling" down the concepts to the age and mental abilities of the children. A recent report by Jeffery Mervis in *Science Now* indicated that students learn "much better through an active, iterative process that involves working through their misconceptions with fellow students and getting immediate feedback from the instructor." There are plenty of misconceptions about science; many students think that science is boring, difficult, and useless. Designing classroom and laboratory activities that will create an atmosphere of self-discovery is very different from the didactic atmosphere that many schools have employed for the last couple of centuries, will help to change this perception.

A research team from the University of British Columbia in Vancouver, Canada, led by the Nobel laureate for physics Carl Wieman, performed an experiment. They had one post-doctorate educator, named Louis Deslauriers, and a graduate student, named Ellen Schelew, use an educational approach called "deliberate practice". This method asked the students to "think like scientists and puzzle out problems during class." The two participants took over a group of introductory physics students for a week. They meet three times within that week for an hour per class. The study shares that:

> *"the results were dramatic… (a)fter the intervention, the students in the deliberate practice section did more than twice as well on a 12-question test of the material as did those in the control section (taught by a tenured physics professor using the standard lecture format). They were also more*

engaged, attendance rose by 20% in the experimental section. In a post study survey they found that nearly all said that they would prefer the entire 15-week course to be taught in a more interactive manner."

Carl Wieman says of the experiment, "It's almost certainly the case that lectures have been ineffective for centuries. But now we've figured out a better way to teach." This new method makes students into active participants throughout the process. Cognitive scientists have also discovered that learning "only happens when you have this intense engagement." Engaging the students through exciting scientific activities will foster self-discovery and help them focus their minds into absorbing the knowledge that science teachers wish to impart to them.

Steven Novella from the *Neurologica* blog gives the following tips to improve science education in the K to 12 level in his post entitled "How to Improve Science Education" from September 5, 2008:

1. *"Teach how we know what we know."* We personally know from experience that just dictating the facts and making the students memorize various scientific theories will not help them retain the knowledge. Novella also says: "Science is a process, not a list of accepted facts… Emphasis should be placed on methodology, what constitutes a proper scientific hypothesis, and how the scientific process unfolds over time."
2. *"Include examples from popular culture and current controversies."* Novella says that educators can use

"popular myths and misconceptions to teach scientific methodology. Science curricula should not shy away from such questions as whether or not bigfoot is likely to exist. Questions that grab the imagination and already exist in popular culture are great fodder for discussion."

3. *"Teach how to access scientific information."* We believe that teaching the students the kind of resources that they can use in order to gain further knowledge will be invaluable in the future. If there are questions that they cannot ask in the classroom, they will have the skills and the knowledge of where to find extra materials.
4. *"Encouraging the use of critical thinking."* Novella adds, "...critical thinking skills should be woven into every part of the science curriculum. Science is, first and foremost, a way of thinking about the world and figuring out how it works." Students should be encouraged to think for themselves and be curious about why some things are and why other things are not.

Novella makes some excellent points on his blog, *Neurologica,* about how we can improve our current education system when it comes to the sciences and we agree with the above tips that he has suggested, but we also know that science education needs more than just these four tips to produces scientifically literal students.

Conclusion

Research indicates that we need a new way to teach science. A way that is not taught with only textbooks and the occasional experiment, but a way that is taught as facts in partnership with practical, hands-on learning. A way that incorporates experts in the field and real-life examples of the science we are trying to teach. We have laid out our plan in this book for a new way to teach science that we believe will lead our students to excel in their science education.

Appendix

What is Notebooking?

We generally recommend notebooking over straight comprehension worksheets during the elementary and early middle school years because we consider it an excellent way of engaging the student with the information he is learning. In notebooking, the student is not merely regurgitating facts; he is thinking over what he has read or heard and responding with what he has found to be meaningful. Notebooking is an extremely effective tool that, over time, will teach the student how to process and release information.

Notebooking has two key components:

1. Material Content
2. Visual Content

The material component of notebooking contains the information the student has learned, while the visual component of notebooking displays a picture of the concept he has studied. Both are equally important since they each engage different parts of the student's brain. Using these two components hand in hand will help to solidify the information into the student's mind as well as train him how to share what he knows. How you arrange the two components of notebooking is up to you and the student.

As the student proceeds through his schooling years, notebooking remains an effective tool that can mature with him. During the elementary years, we compared the student to an empty bucket that is waiting to be filled with information. Notebooking is the go-to tool you can use to verify that he has placed a piece of information into his bucket. As the student moves into the middle school years, he has a filled bucket of information, but the material still needs to be added to and organized during these years. Notebooking remains a handy method that you can use to help him arrange the data that is swimming around in his head.

In first grade, notebooking can be very basic. For example, after you have read about the animal you are studying, simply ask, "What is one thing that you learned about the animal? or "What do you like best about the animal?" The student's answers will be simple, something like:

Lions roar.

OR

Male lions have a mane.

Sample Notebooking Page From a 1st Grader

Once you have written the answer down for him, you can have him add a picture of the animal he studied. This can be a picture that you have printed or one that the student has drawn and colored.

Around the third grade, notebooking begins to increase in difficulty. At this age, the student should be able to begin writing his own answers, or narrations, for his notebooking pages. His narrations can be in paragraph form or in list form. You can still ask him broad questions like, "What are several things that you learned about the alkali metals?" To which he may answer

> *They are soft metals in your house. They are very reactive. They have low melting points.*

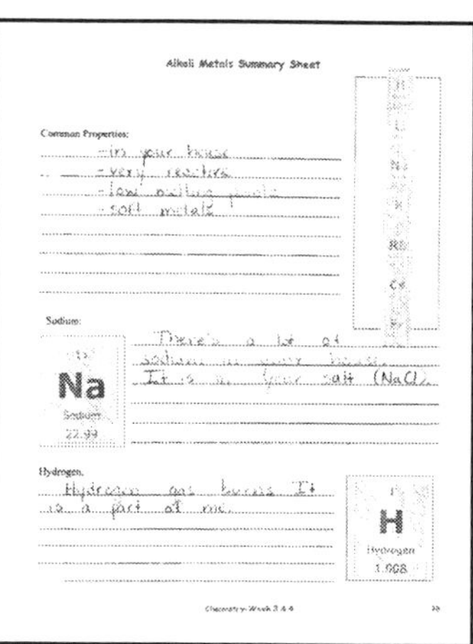

Sample Notebooking Page From a 3rd Grader

Once he has finished writing, you can have him paste in a picture of the alkali metal group. It is important to note that as the student gets older, you are still making use of the visual component. You don't want to neglect this step because the pictures will serve as visual markers of the concept in the student's mind.

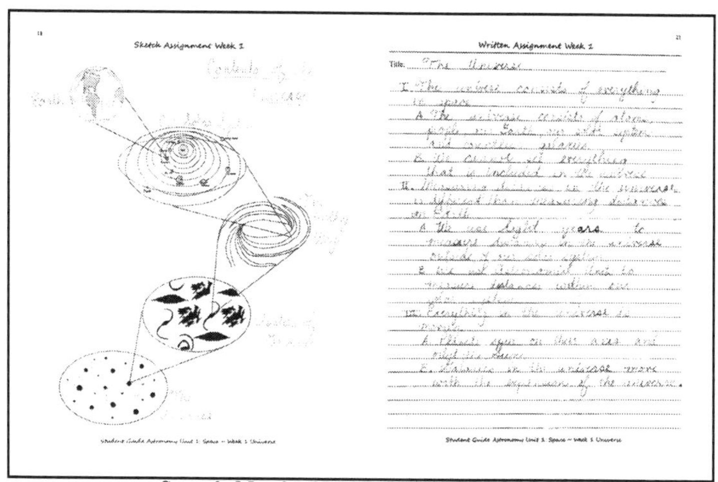

Sample Notebooking Page From a 6th Grader

As the student gets older, his notebooking will reflect his growing intellect. His material component will include longer paragraphs or outlines and his visual component will turn into sketches or diagrams. In addition, instead of asking simple broad questions about his favorite parts, you will ask leading questions that will help him to pull out the most important information from what he reads.

So, in sixth grade the student will be coloring and labeling a sketch that comes from what he studied as well as crafting an outline or a report from what he read. This type of notebooking looks a bit different than it did in earlier years, but it is still equally as effective.

The material and visual components of notebooking serve as two interlocked parts that will help you to firmly affix key pieces of information into the student's brain. We recommend that notebooking be an integral feature of any science curriculum because we believe that it is a vital part of an excellent science education.

What is Nature Study?

Nature study is a style of educating that searches for the principles of science in nature. It became a popular educational movement in the early 20th century and it is currently regaining popularity once again among homeschoolers. Anna Botsford Comstock, one of the pioneers of the movement, shares the following about nature study in her book, *Handbook of Nature Study:*

> *"**Nature study cultivates the child's imagination,** since there are so many wonderful and true stories that he may read with his own eyes, which affect his imagination as much as does fairy lore, at the same time nature study cultivates in him a perception and a regard for what is true, and the power to express it...**Nature study gives the child practical and helpful knowledge.** It makes him familiar with nature's ways and forces, so that he is not so helpless in the presence of natural misfortune and disasters."*

Nature study awakens the scientific side of the brain in the same way that a good book can awaken the imagination. It helps the student to see science face to face, to understand the processes going on around him every day and to develop in him a hunger to learn more.

From the dawn of the ages, science has been found in nature. The study of science began with philosophers observing things in nature that made them question why. As they began to seek for the answers to their questions, the study of science was born. Today, scientists still employ the power of observation daily, so it is important that we impart this skill to our budding scientist. Nature study can be one of the tools we use to train him how to observe because it teaches the student to slow down and really see the science surrounding him.

Nature study has two key components:

1. The Nature Walk
2. The Nature Journal

Nature study begins with the nature walk. This is a time set aside to go out and explore the world around you. You can do this by taking a simple stroll through the area you live in or by hiking along a nearby trail. Usually these nature walks are spent looking for something specific as you go along, kind of a "walk with a purpose."

As you ramble along the path, you will take the time to point out the focus of your nature study. So, if you are studying oak trees, you will point out any you see along the way. Then take a moment to stop by one of the oak trees so that you can discuss the shape of its leaves, the texture of its bark and any acorns or seeds you may find on the ground. Allow the student to ask any questions he has and provide him with any important information about the oak tree that you would like for him to know. You can have the student collect samples, such as a leaf or an acorn, make a rubbing of the bark and then step back to get a picture of the whole tree. As you continue on your walk, point out any additional oak trees and compare them to the tree that you stopped to study. You can also compare them to the other types of trees along the path.

The second component of a nature study is the nature journal. You can pick a spot along your journey to sit down and write about your experience or you can wait until you get home to have the student add an entry to his nature journal. Either way, the nature journal should be a personal record of what the student has learned. In it he should feel free to glue in pictures or samples, to draw what he has seen and to record his thoughts. I also recommend that every student records at least one scientific fact about the subject being studied and the date with each entry.

Sample Nature Journal Entry from a 2nd Grader

The nature walk and nature journal work together to show the student the wonders of science in the world around him. When used in conjunction, they create a full nature study, which can be an exceptionally effective tool for science education.

What are Lapbooks?

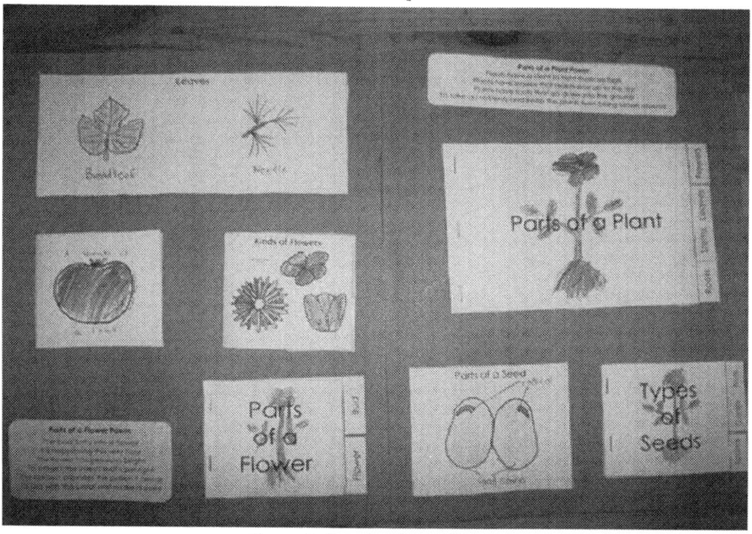

Sample Lapbook on Plants

Lapbooks are educational scrapbooks that fit into the lap of a student. Typically they are a collection of related mini-books on a certain subject that have been glued into a file folder for easy viewing, but they can also include pictures or projects that the student has completed. In the same way that notebooking does not require regurgitation of facts; lapbooking causes the student to interact with the materials instead of just responding to comprehension questions.

Lapbooks are extremely versatile because they can be used in conjunction with any subject the student is learning about. They are excellent tools to use with elementary student as a way of reinforcing what he is learning because this age group tends to prefer a more creative format of notebooking.

The heartbeat of the lapbook is the mini-books that are placed inside. Each of these booklets contains information on topics related to the main subject of the lapbook. They can be in a variety of shapes and sizes, but the cover should have a picture related to the subject as well as a title. The interior of each booklet should contain several sentences detailing what the student has learned about the topic in his own words. The mini-books will each pertain to different sub-topics of the main topic. So for instance, if your main topic was matter, your related mini-books could be on the states of matter, changes in stage, solids, liquids, gases and more.

Lapbooks serve as beautiful scrapbooks that the student can continue to learn from for years to come, which makes them a beneficial addition to a student's science education.

What is a Living Book?

An Example of a Living Book Related to Science

A living book engages the reader and draws him into learning more about a subject. It is typically narrative in style and written by someone with a passion for the subject or by someone who has experienced the story first hand. The author pulls the reader into the story and presents facts in such a way that the student hardly realizes that he is learning.

In some ways, the living book is a superior learning tool when compared to the typical textbook or encyclopedia. This is because these reference works cover a broad range of topics that are shared in a dry, but systematic way. The student can have difficulty engaging with the subject and thus will be less apt to remember what he has read. On the other hand, the living book draws the reader in and presents the facts as part of a story line, which leads to greater retention of the material.

However, living books are somewhat limited because they typically focus on one specific topic, which is thoroughly explored. This can be a positive feature, but it also has one major drawback. In a living book, you must read the story as a whole to engage with the material, while in a reference book you can quickly access the information you need. So, whether you choose a living book or reference work will depend on the time you have to devote to your study as well as how much material you would like to cover in that time.

When choosing living books for science, make sure you look for books that will draw a student in, but are still educationally sound. You don't want the student's first impression of a subject to be polluted with false material. When looking for books for the youngest student, make sure that the book has a fair amount of quality illustrations that will hold his attention as you read the book to him.

The entire purpose of using living books in science education is to engage the student with the subject he is studying so that he will be more apt to remember what he has studied.

What Should a Lab Report Include?

It is very important that all high school students begin to understand the process of writing scientific laboratory reports. Learning technical writing is a skill that must be practiced in order to become proficient. Having the student write laboratory reports also helps him to think critically as he analyzes the data that he observed in his experiments and gives the student a good idea of the basics of scientific writing that will help to prepare him for college level work.

The Components

Although there are some variations to a scientific lab report depending on the discipline you are writing for, each contains the same basic components. Below is a general outline of the components of a scientific lab report.

1. Title
2. Abstract
3. Introduction
4. Materials and Procedure
5. Results (includes observations and data)
6. Conclusion
7. Works Cited

Each section of the scientific lab report should include specific information. The following is an explanation of what each section should have.

Title

This section should summarize the scientific experiment

in 10 words or less and use key words in the report.

Abstract

An abstract gives a one paragraph synopsis of the research that the scientist has found. Generally the scientist will write a full research report that is associated with his experiment and the abstract can let the reader know if reading the full report would be beneficial. It should contain the data and conclusions of the report in 200 words or less. If the student has not completed an associated research report, he can use this to summarize his research.

Introduction

The introduction section of the report is designed to give the reader the basis for the report, the reason why it was completed and the background about what is already known about the experiment. Usually this section answers, "Why did we do this study?", "What information was known before we began the experiment?" and "What is the purpose of the study?"

Materials and Procedures

This section tells the reader what specific equipment and supplies were used in the experiment. You will also need to describe how the equipment and supplies were used. It is also important to describe when and where the experiment was conducted. The purpose of this information is to allow any person that reads the scientific report to be able to replicate the results. Without being able to replicate the results the data obtained is unconfirmable.

Results

The main purpose of the results section is to present the

data that was obtained during the experiment. It is important that you present just the data and avoid analyzing or making conclusions about the data as there is another section for this. The results section is also the place to include tables, graphs and data tables. These should be easily understood by the average reader and well-labeled.

Conclusion

In the conclusion section one should focus on analyzing the data. Don't just reiterate the data but rather draw conclusions from the data. It is permissible to draw speculative conclusions, just remember to state that is what you are doing. In this section, you should also discuss if the hypothesis was confirmed or invalidated and if further experimentation should be carried out to refine your hypothesis. You must also discuss any errors that occurred during the lab, these are usually not human errors but rather systematic errors that occur while conducting the experiment. Systematic errors can be caused by improper calibration of equipment, changes in the environment and estimation errors. In the conclusion it is important to present these errors that may have occurred in your experiment. This section is usually written in the third person, passive past tense.

Works Cited

This is the listing of the research materials used in the report to give background or to help corroborate the data. All outside sources must be cited. The format you will use for the works cited (i.e. MLA, APA, Turabian etc.) will depend upon the scientific field that this lab report relates to.

The Style

When formatting the lab reports there are some basic guidelines about style that each student should know.

1. They should have a 1 inch margin and use 12pt New Times Roman font.
2. All chemical formulas should be formatted properly (i.e. Cl_2, H_2O).
3. Chemical structures that are drawn should be neat and easily readable.

You also need to pay attention to formatting, spelling and the overall look of the lab report.

More Information on Lab Reports

We recommend two books for the student to better prepare for writing scientific papers and lab reports. These books cover in greater detail what we covered in the previous section and they can be easily obtained from Amazon or your university library.

1. *A Student Handbook For Writing In Biology,* written by Karin Kinsely
2. *Making Sense: Life Sciences: A Student's Guide to Research and Writing,* written by Margot Northey

Bibliography

Augustine, Norm. "Danger: America Is Losing Its Edge In Innovation". *Forbes Magazine*. Forbes, Jan. 11, 2011. http://www.forbes.com/sites/ciocentral/2011/01/20/danger-america-is-losing-its-edge-in-innovation/

Bauer, Susan Wise. "Science in the Classical Curriculum" *Peace Hill Press*, Audio Lecture. http://peacehillpress.com

Ben-Chaim, David, Salit Ron, and Uri Zoller. "The Disposition of Eleventh-Grade Science Students toward Critical Thinking." *Journal of Science Education and Technology* 9.2 (2000): 149-59. Print.

Bredderman, Ted. "Effects of Activity-Based Elementary Science on Student Outcomes: A Quantitative Synthesis." *Review of Educational Research* 53.4 (1983): 499. Print.

Bybee, Rodger W., and Cherilynn A. Morrow. "Improving Science Education: The Role of Scientists." *Nasa*.gov. Nasa.gov, 2005. Web. 15 June 2012. http://sunearthday.nasa.gov/2005/materials/BMroles.pdf.

California. California Department of Education. *Science Content Standards for California Public Schools Kindergarten Through Grade Twelve*. Ed. Sheila Bruton Bruton. By Faye Ong. California Department of Education, 2000. Web. 16 June 2012.

Chang, Chun-Yen, Ting-Kuang Yeh, Chun-Yen Lin, Yueh-Hsia Chang, and Chia-Li D. Chen. "The Impact of Congruency Between Preferred and Actual Learning Environments on Tenth Graders' Science Literacy in Taiwan." *Journal of Science Education and Technology* 19.4 (2010): 332-40. Print.

Comstock, Anna Botsford. *Handbook of Nature Study*. Ithaca: Comstock Publishing Associates, 1967, Print.

DeBoer, George E. "Scientific Literacy: Another Look at Its Historical and Contemporary Meanings and Its Relationship to Science Education Reform." *Journal Of Research In Science Teaching* 37.6 (2000): 582-601. Print.

Deslauriers, Louis, Schelew, Ellen and Wieman, Carl. "Improved Learning in a Large-Enrollment Physics Class." *Science* 13 May 2011: 332 (6031), 862-864. Print.

Fraser, Barry J., Kenneth George Tobin, and Campbell J. McRobbie, eds. *Second International Handbook of Science Education*. 24th ed. Vol. 1. Dordrecht: Springer, 2010. Print.

Freedman, David. "Impatient Futurist: Science Finds a Better Way to Teach Science." Discover Magazine 01 Dec. 2011: n. pag. *Discover Magazine*. Discover, 17 Jan. 2012. Web. 17 June 2012. http://discovermagazine.com/2011/dec/16-impatient-futurist-science-finds-better-way-to-teach.

Hancock, LynNell. "Smithsonian Magazine." *Smithsonian Magazine*. Smithsonian, Sept. 2011. Web. 06 July 2012. http://www.smithsonianmag.com/people-places/Why-Are-Finlands-Schools-Successful.html.

Harty, Sheila. "Project 2061: Systemic Reform of K-12 Education for Science Literacy." *Journal of Science Education and Technology* 2.3 (1993): 505-07. Print.

Klahr, David, and Junlei Li. "Cognitive Research and Elementary Science Instruction: From the Laboratory, to the Classroom, and Back." *Journal of Science Education and Technology* 14.2 (2005): 217-38. Print.

McMillan, Victoria E. *Writing Papers in the Biological Sciences*. Boston, Mass. [u.a.: Bedford/St. Martins], 2006. Print.

Mervis, Jeffrey. "A Better Way to Teach?" *ScienceNOW*. AAAS, 12 May 2011. Web. 01 July 2012. http://news.sciencemag.org/sciencenow/2011/05/a-better-way-to-teach.html.

Mooney, Chris. "What's Wrong with U.S. Science Education?" *Science Progress*. Center for American Progress, 5 Aug. 2005. Web. 06 July 2012. http://www.scienceprogress.org/.

Mooney, Chris & Kirshenbaum, Sheril . *Unscientific America: How Scientific Illiteracy Threatens our Future*. New York, New York. Basic Books, 2009. Print

Moore, Randy. "What's Wrong with Science Education & How Do We Fix It?" *The American Biology Teacher* 2.6 (1990): 330+. JSTOR. University of California Press. Web. 15 June 2012. http://www.jstor.org/stable/4449128.

Nanjundiah, Sadanand. "The Training of Science Teachers in the Peoples Republic of China." *Journal of Science Education and Technology* 5.2 (1996): 161-65. Print.

"A Nation at Risk: The Imperative for Educational Reform." Rep. N.p.: Nation at Risk: The Imperative for Educational Reform, 2005. A Nation at Risk: The Imperative for Educational Reform. US Department of Education, 1983. Web. 15 June 2012.

National Science Education Standards. *Inquiry and the National Science Education Standards: A Guide for Teaching and Learning*. Board on Science Education, 2000.

National Research Council. "Rising Above the Gathering Storm, Revisited: Rapidly Approaching Category 5. Washington, DC:" *The National Academies Press*, 2010.

Northey, Margot, and David B. Knight. *Making Sense: A Student's Guide to Research and Writing : Geography & Environmental Sciences*. Don Mills, Ont.: Oxford UP, 2001. Print.

Northey, Margot, and P. Von. Aderkas. *Making Sense: Life*

Sciences : A Student's Guide to Research and Writing. Don Mills, Ont.: Oxford UP, 2011. Print.

Novella, Steven. "How To Improve Science Education." Web log post. *NeuroLogica*. N.p., 5 Sept. 2008. Web. 12 June 2012. http://theness.com/neurologicablog.

Partanen, Anu. "What Americans Keep Ignoring About Finland's School Success." *The Atlantic*. Atlantic Monthly Group, 29 Dec. 2011. Web. 06 July 2012. http://www.theatlantic.com/national/archive/2011/12/what-americans-keep-ignoring-about-finlands-school-success/250564/.

"Researchers Find Systemic Problems in U.S. Mathematics and Science Education." *New Educator*. Michigan State University, Fall 2000. Web. 06 July 2012. http://www.educ.msu.edu/neweducator/fall00/Timss.htm.

Ruby, Allen. "Hands-on Science and Student Achievement." *Diss. Rand Graduate School*, 2001. (2001): 1-272. Hands-on Science and Student Achievement. Rand Corporation, 2001. Web. 15 June 2012.

Sahlberg, Pasi, and Andy Hargreaves. *Finnish Lessons: What Can the World Learn from Educational Change in Finland?* New York: Teachers College, 2011. Print.

Scientific Style and Format: The CBE Manual for Authors, Editors and Publishers. Bethesda, MD: Council of Biology Editors, 1994. Print.

Tobias, Sheila. "What Makes Science Hard? A Karplus Lecture." *Journal of Science Education and Technology* 2.1 (1993): 297-304. Print.

Vogt, Kimberly J. *The Effects of Hands-On Activities on Student Understanding and Motivation in Science*. Diss. University of Dayton, 2005. N.p.: n.p., n.d. Print.

16289555R00081

Made in the USA
Charleston, SC
13 December 2012